TECHNO MIC
PUBLISHING C

R E V I S E D

Primer on Composite Materials: ANALYSIS

J. C. HALPIN
Air Force Materials Center
Wright-Patterson AFB, Ohio

Materials Research Laboratory
Washington University
St. Louis, Missouri

851 New Holland Avenue
Box 3535
Lancaster, Pennsylvania 17604

Printed in U.S.A.
Library of Congress Card No. 84-50921
ISBN 87762-349-X

C O N T E N T S

P R E F A C E

The first edition of this Primer was conceived in 1968 as an instrument for the training of engineer and structural designers. For more than a decade, this Primer has been used as the introductory text, both in industry and university courses.

The material in the first edition was developed in lecture courses at Washington University, and in the training of personnel at General Dynamics, Fort Worth Texas, and Lockheed Georgia, Marietta, Georgia. Accordingly, the first edition served as a text for the novice in the field of mechanical properties of anisotropic layered systems, and prepared workers in the field for intelligent research in the area. Both qualitative and quantitative aspects of the analysis of composite materials were analyzed in the first edition as well as in this revised edition.

In response to a continuing demand for the Primer, it has been revised and expanded. The expansion will result in three individual volumes:

1. Analysis
2. Fracture Control and Design
3. Materials and Processes

Each volume will be of similar size to the original "Primer on Composite Materials: Analysis"; will be technically interlocking; and suitable for a three quarter academic or industrial training program.

The major thrusts of the original Primer "Analysis" have been preserved in this revision. Increased emphasis on the "Strength of Laminated Composites" resulted in a new Chapter 4. An expanded treatment of the dimensional consequences of thermal and environmental induced deformations has been included. This addition reflects the increasing role of laminated composite analysis in electronic packaging, printed wiring board fatigue analysis, prosthetic devices, space communications, etc. Additional material has also been added regarding laminated beams and shells in Chapter 5.

As with the first edition, many topics in this revised Primer are developed to a high degree of sophistication, thereby providing a worthwhile source of information on for the researcher who is already acquainted with the field.

Dayton, Ohio
April 1984

JOHN C. HALPIN

1
Introduction

1.1 THE NATURE AND SCOPE OF COMPOSITE MATERIALS

WE SHALL BE PRIMARILY CONCERNED WITH A CLASS OF COMPLEX materials, composites, in which nonwoven fibers are deliberately oriented in a matrix in such a way as to increase its structural efficiency. Current developments are pointed toward combinations of unusually strong, high modulus fibers and organic, ceramic, or metal matrices. Such materials promise to be far more efficient than any structural materials known previously, and to extend the temperature-performance spectrum into regions unattainable only a few years ago.

Because of the variety of combinations and arrangements of fibers and matrices combined with the concept of lamination, designers have today greatly increased opportunities for tailoring structures and/or materials to meet systems of forces and changing environments. Achieving this goal necessitates new techniques in design and manufacture and the compilation of vast amounts of information on the properties of a whole new family of materials.

This situation exists today despite the fact that the concept of fiber reinforcement was known and employed in ancient times. As an example, straw was used by the Israelites in the manufacture of bricks in 800 BC. It was realized at that time, that the addition of chopped straw or other fibrous vegetable matter was an improvement from several points of view. It assisted the evaporation of moisture from the interior, and helped to distribute the cracks which were formed more evenly, thereby increasing the "structural performance" of the mud brick. A short time thereafter, the Mongol bow was constructed of a composite of animal tendons, wood, and silk bonded together with an adhesive. Still later, laminated structure appeared in the form of the Damascus gun barrels and the Japanese ceremonial swords.

The first "high performance" composite material is as old as man himself, for it is the human body: the bones; muscle tissue which is a multidirectional fibrous laminate; etc. The most widely used high performance composite of commercial importance today, and the forerunner of modern composite material science, is the pneumatic tire. In fact, it is rather interesting to note that modern technology is, in many ways, repeating the practices of the past. For not only are we now using fibers for strength and stiffness reinforcement, but, in addition, modern fabricators are employing resistant heating of the reinforcements themselves to aid in the rapid curing of composites without inducing high internal cure stresses. Shades of the Israelites conducting moisture from mud blocks with straw.

In this monograph an elementary but detailed development of the new design procedures for modern fibrous composites shall be developed. Our emphasis is on stiffness properties and/or calculations. In order that we might illustrate an elementary design procedure it will also be necessary for us to discuss briefly a failure or strength criterion. This is accomplished by adopting the simplest criterion possible. In a subsequent volume, a more advanced, accurate criterion shall be developed. However, the procedures by which one employs a failure criterion will be the same as oulined here: the simple criterion being adopted here solely for the purpose of analytical ease.

1.2 COMPOSITE MATERIALS TERMINOLOGY

The two terms which describe the engineering materials with which the materials scientist, analyst, or designer normally works (i.e., metals, thermosetting, or thermoplastic organic polymers) are homogeneous and isotropic. In contrast, a laminated composite material may be described by any of the following terms: homogeneous orthotropic; homogeneous anisotropic; heterogeneous orthotropic; heterogeneous anisotropic; and quasi-isotropic. The materials scientist, analyst, and designer must become familiar with these terms and with the concepts on which they are based in order to become proficient in the use of composites and to be able to communicate regarding composite materials. These terms and other terms pertinent to composites will be defined in this section.

Generally speaking, a composite material is a material with several distinct phases present. Normally, the composite consists of a reinforcing material (fiber, whisker) supported in a binder or matrix material. The reinforcing material is normally the load carrying medium in the material, and the matrix serves as a carrier, protector, and load splicing medium around the reinforcement.

One of the first applications of composite materials occurred in the aircraft industry when cloth fiber glass laminates were utilized for secondary structural applications such as radomes. Composite materials were first used in primary structural applications when filament-wound (glass) pressure vessels were developed for the Polaris missile engines. The birth of so-called advanced composites began with the fabrication of boron filaments in the early 60's. These continuous filaments are considerably stiffer and stronger than the glass filaments which preceded them; hence, the name advanced composite materials evolved.

The term filamentary composites is often used to differentiate between continuous filament composites and short filament or whisker reinforced composites. A filamentary composite is normally made up of several plies or laminae. One lamina of a filamentary composite consists of one row of parallel filaments surrounded by the matrix, Figure 1-1. The lamina are stacked with various orientations of the filament direction between lamina to obtain a laminate which has the desired stiffness or strength properties.

As previously stated, the two terms which describe the engineering materials (i.e., metals) with which the aircraft stress analyst or designer normally works are homogeneous and isotropic. The term homogeneous means uniform. That is, as one moves from point to point in the material, the material properties (i.e., stiffness, thermal conductivity, thermal coefficient of expansion, etc.) remain constant. The material properties are not a function of *position* in the material. The term isotropic indicates that the material properties at a point in the body are not a function of orientation. In other words, the material properties remain constant regardless of the reference coordinate system at a point. As a result, the material properties are constant in *any* plane which passes through a point in the material. Therefore, for an isotropic material, *all* planes which pass through a point in the material are planes of material property symmetry.

In an orthotropic material, only *three* mutually perpendicular planes of material property symmetry may be passed through a point. For an anisotropic material, there are *no* planes of material property symmetry which pass through a point. Therefore, for an orthotropic and for an anisotropic material, one would expect the material constants at a point to change as one rotated the coordinate system at the point. Isotropic and orthotropic materials are specialized classes of materials which have a higher degree of symmetry than isotropic materials.

A material may be isotropic, orthotropic, or anisotropic and either be homogeneous or heterogeneous. If a material is homogeneous, the material

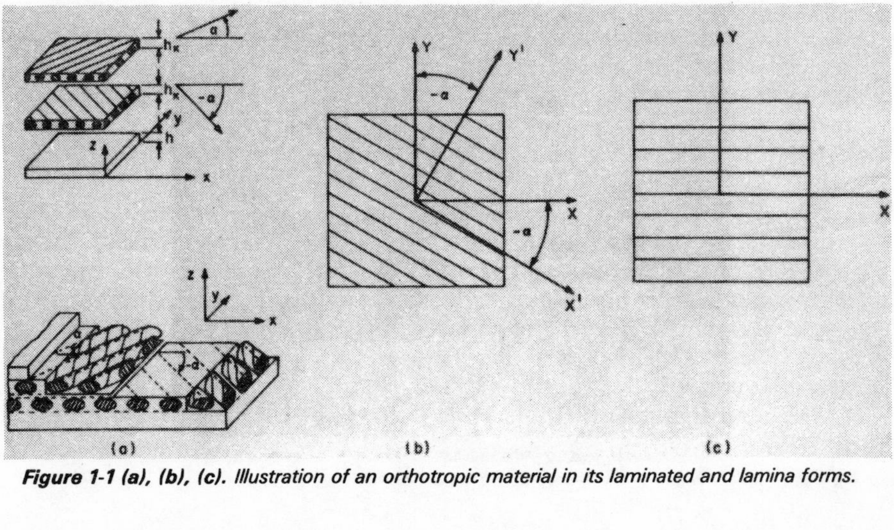

Figure 1-1 (a), (b), (c). *Illustration of an orthotropic material in its laminated and lamina forms.*

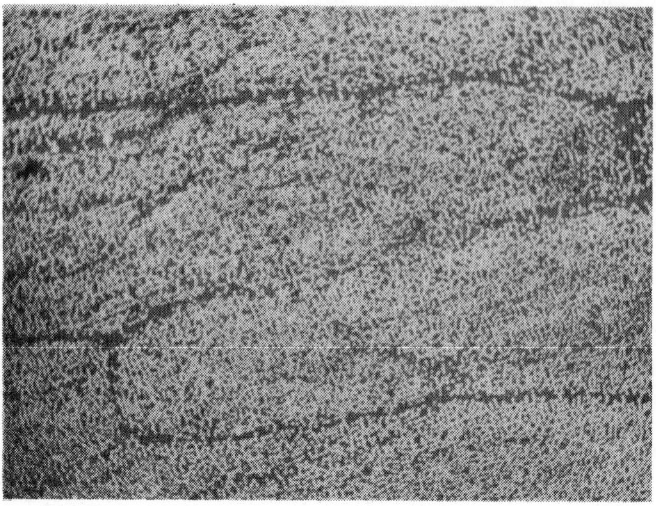

Figure 1-1 (d). Photomicrograph of a lamina of a graphite/epoxy material (100 ×).

properties do not change from point to point in the body although they may change with a coordinate rotation at the points (orthotropic or anisotropic). As long as the material properties are the same for the same coordinate position from point to point in a body, the material is homogeneous. If this is not true, the material is said to be heterogeneous.

By referring to Figure 1-1, it is evident that the individual ply or lamina of a filamentary composite has three mutually perpendicular planes of symmetry for the material constants. The intersection of the three planes forms the axes

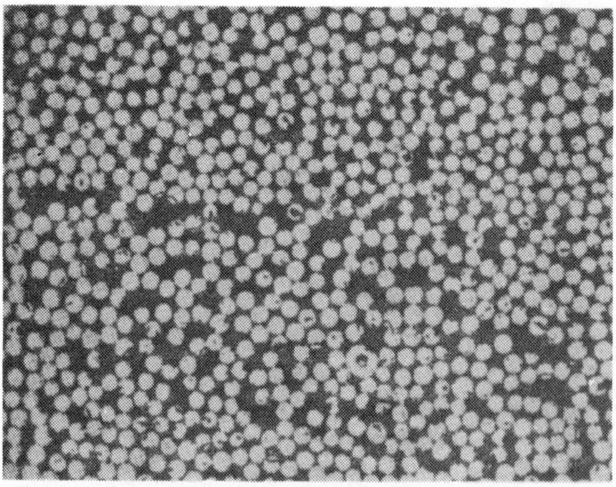

Figure 1-1 (e). Enlarged photomicrograph of a fiber bonded in a graphite/epoxy material (500 ×).

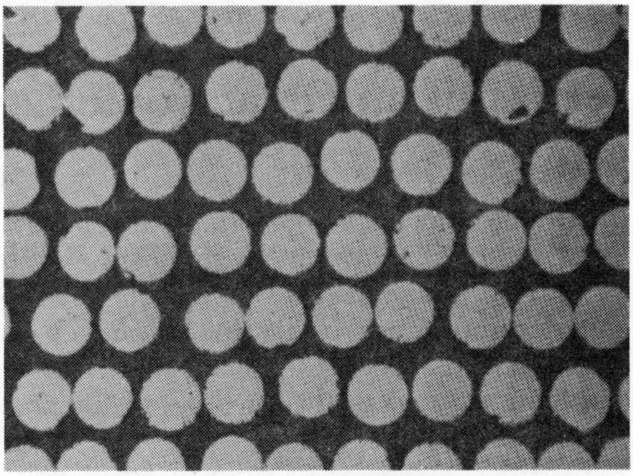

Figure 1-1 (f). *Boron laminate (100 ×).*

of the coordinate system shown. Since the lamina has three perpendicular planes of symmetry, it can be considered to be an orthotropic material on the macroscopic level. Also, the lamina may be considered homogeneous on the macro scale. The set of axes which are parallel and perpendicular to the filament direction are termed the lamina principal axes.

In the literature on composite materials, one often encounters the term constitutive equation. It is a term borrowed from the science of Rheology. The term constitutive refers to the stress-strain or Hooke's law relationships for a material because the stress-strain relations actually describe the mechanical constitution of the material.

The term quasi-isotropic is used quite often in the literature on composite material. The term is used to describe a particular type of laminate—a laminate which has essentially isotropic stiffnesses and perhaps strength. The simplest quasi-isotropic laminate is a three ply laminate with a $0°$, $\pm 60°$ orientation. The next possible quasi-isotropic laminate is the four ply $0°$, $\pm 45°$, $90°$ laminate. Of course, the more plies in the laminate, the less is the angle between the principal axes of each lamina, and the laminate will better simulate an isotropic condition.

2
Properties of an Orthotropic Lamina

IN THIS SECTION, THE MECHANICS GOVERNING THE BEHAVIOR OF THE INdividual lamina or ply of a laminated composite will be developed. The discussion will be generalized to include laminae which are isotropic, orthotropic, or anisotropic in nature. The development will first begin with the general concepts of stress and strain and the generalized Hooke's law and then proceed to the specifics of the mechanics of the lamina.

2.1 STRESS AND STRAIN

In this section, a brief review of the fundamental principles of stress and strain at a point will be made in order to form a firm base for understanding the mechanics of the lamina.

As pointed out in Appendix A, stress and strain are physical entities which can be classified as second rank tensors. Because they are second rank tensors, the relations which transform one stress or strain state to a new coordinate system are well known. The graphical form of these transformation relations is the familiar Mohr's cycle.

The analytical expressions for the stress transformation relations can be derived from considerations of equilibrium on a small element. By restricting ourselves to a two dimensional stress state, Figure 2-1, a summation of forces in the "one" direction on free body *A* gives the following equation:

$$\sigma_1 dA - \sigma_x(\cos\theta\, dA)(\cos\theta) - \sigma_y(\sin\theta\, dA)(\sin\theta) \\ - \tau_{xy}(\sin\theta\, dA)(\cos\theta) - \tau_{xy}(\cos\theta\, dA)(\sin\theta) = 0$$

By rearranging and simplifiying the above equation, a relationship is obtained between the normal stress, σ_1, and the stress components in the *x-y* coordinate system.

$$\sigma_1 = \sigma_x(\cos^2\theta) + \sigma_y(\sin^2\theta) + \tau_{xy}(2\sin\theta\cos\theta) \quad (2\text{-}1)$$

A summation of forces in the "two" direction on free body *A* gives the following:

$$\tau_{12} dA + \sigma_x(\cos\theta\, dA)(\sin\theta) - \sigma_y(\sin\theta\, dA)(\cos\theta) \\ - \tau_{xy}(\cos\theta\, dA)(\cos\theta) + \tau_{xy}(\sin\theta\, dA)(\sin\theta) = 0$$

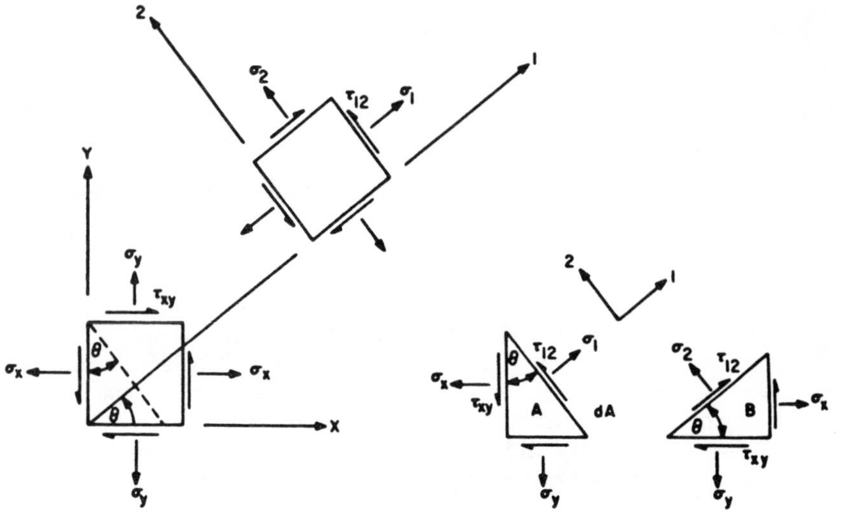

Figure 2-1. Stress coordinate rotation.

Simplifying the above equation, the following is obtained:

$$\tau_{12} = -\sigma_x(\sin\theta\cos\theta) + \sigma_y(\sin\theta\cos\theta) + \tau_{xy}(\cos^2\theta - \sin^2\theta) \quad (2\text{-}2)$$

A summation of forces in the "two" direction on free body B gives the following:

$$\sigma_2 dA - \sigma_x(\sin\theta\, dA)(\sin\theta) + \tau_{xy}(\sin\theta\, dA)(\cos\theta) + \tau_{xy}(\cos\theta\, dA)(\sin\theta) - \sigma_y(\cos\theta\, dA)(\cos\theta) = 0$$

By simplifying the above equation, a relationship is obtained for the remaining normal stress, σ_2, in terms of the stress components in the x-y coordinate system.

$$\sigma_2 = \sigma_x(\sin^2\theta) + \sigma_y(\cos^2\theta) - \tau_{xy}(2\sin\theta\cos\theta) \quad (2\text{-}3)$$

A summation of forces in the "one" direction on free body B gives the same relations for shear stress as equation 2-2.

When the strain components, ε_x, ε_y, and γ_{xy} are known at a point, the strain components may be determined for any other orientation of the reference axes by obtaining the strain transformation relations. The strain transformation relations may be obtained from the deformation of a line segment, Figure 2-2. The line segment, PQ, which is of infinitesimal length, is extended or contracted, rotated, and translated during the deformation to the final position, $P'Q'$. If the line $P'Q'$ is translated so that P' coincides with P, then the rleative displacements of the point Q with respect so the point P after deformation are given by the following:

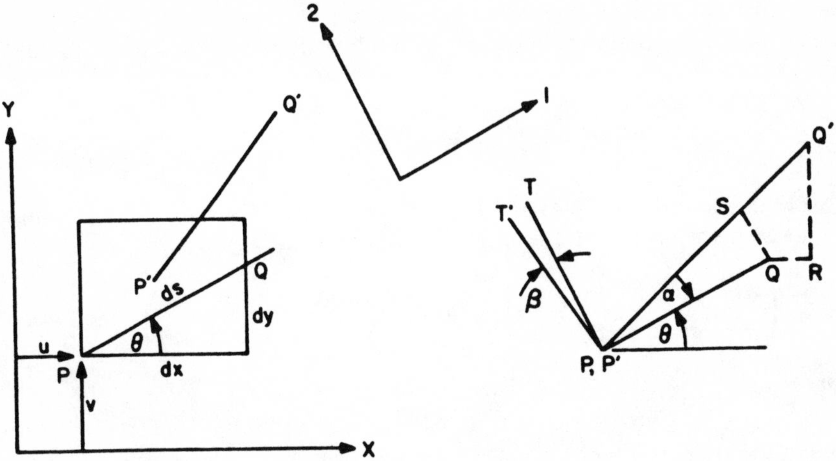

Figure 2-2. *Strain coordinate rotation.*

$$QR = \frac{\Delta u}{\Delta x}\Delta x + \frac{\Delta u}{\Delta y}\Delta y$$

$$Q'R = \frac{\Delta v}{\Delta x}\Delta x + \frac{\Delta v}{\Delta y}\Delta y$$

Since

$$u = f(x,y)$$
$$v = f(x,y)$$

and the line segment PQ is infinitesimal (Δx and $\Delta y \to 0$), then

$$QR = \frac{\partial u}{\partial x}dx + \frac{\partial u}{\partial y}dy$$
$$Q'R = \frac{\partial v}{\partial x}dx + \frac{\partial v}{\partial y}dy \qquad (2\text{-}4)$$

Now, in order to determine the *strains* referred to the 1-2 axes, the relative displacements of point Q with respect to point P must be determined parallel and perpendicular to the line segment PQ. From elementary trigonometry, the relative displacements can be determined to be:

$$QS = Q'R\cos\theta - QR\sin\theta$$
$$SQ' = QR\cos\theta + Q'R\sin\theta \qquad (2\text{-}5)$$

where the small angle α is ignored in comparison with the angle θ. The normal strain which occurs in the direction of the line segment PQ is given by the following:

$$\varepsilon_1 = \frac{SQ'}{PQ} \tag{2-6}$$

Substituting equations (2-4) and (2-5) into equation (2-6) gives the following expression for normal strain in the "one" direction:

$$\varepsilon_1 = \left(\frac{\partial u}{\partial x}\frac{dx}{ds} + \frac{\partial u}{\partial y}\frac{dy}{ds}\right)\cos\theta + \left(\frac{\partial v}{\partial x}\frac{dx}{ds} + \frac{\partial v}{\partial y}\frac{dy}{ds}\right)\sin\theta$$

or

$$\varepsilon_1 = \left(\frac{\partial u}{\partial x}\right)\cos^2\theta + \left(\frac{\partial u}{\partial y} + \frac{\partial v}{\partial x}\right)\sin\theta\cos\theta + \left(\frac{\partial v}{\partial y}\right)\sin^2\theta \tag{2-7}$$

With reference to Figure 2-3, it may be shown that the terms in parentheses are the normal and shear strains which are referred to the x-y axes.

$$\varepsilon_x = \frac{\left(u + \frac{\partial u}{\partial x}dx\right) - u}{dx}$$

$$\varepsilon_x = \frac{\partial u}{\partial x} \tag{2-8}$$

and similarly

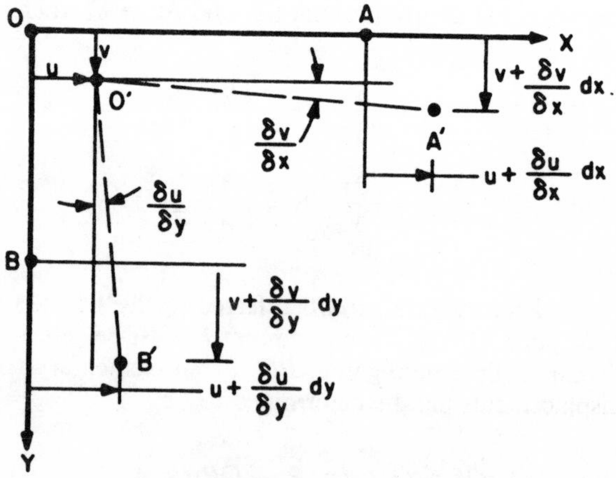

Figure 2-3. Strain-deformation relations.

$$\varepsilon_y = \frac{\partial v}{\partial y}$$

Since the shear strain is the change in the right angle *AOB*, then

$$\gamma_{xy} = \frac{\partial u}{\partial y} + \frac{\partial v}{\partial x}$$

Now, Equation (2-7) can be written in a form which expresses the normal strain, in the "one" direction in terms of the strains in the *x-y* coordinate system:

$$\varepsilon_1 = \varepsilon_x \cos^2\theta + \varepsilon_y \sin^2\theta + \gamma_{xy}\sin\theta\cos\theta \qquad (2\text{-}9)$$

The expression for the normal strain transverse to the line segment *PQ* can be obtained from equation (2-9) by substituting for θ the angle $\theta + 90°$. This gives:

$$\varepsilon_2 = \varepsilon_x \sin^2\theta + \varepsilon_y \cos^2\theta - \gamma_{xy}\sin\theta\cos\theta \qquad (2\text{-}10)$$

In order to determine the shear strain transformations, the change in the 90° angle between line segments *PQ* and *PT* must be determined. The angle α through which *PQ* is rotated is given by the familiar trigonometric relation for small angles:

$$\alpha = Arc\ Length/Radius$$
$$\alpha = QS/PQ$$

By substituting from equations (2-4) and (2-5), the following is obtained:

$$\alpha = \left(\frac{\partial v}{\partial x}\frac{dx}{ds} + \frac{\partial v}{\partial y}\frac{dy}{ds}\right)\cos\theta - \left(\frac{\partial u}{\partial x}\frac{dx}{ds} + \frac{\partial u}{\partial y}\frac{dy}{ds}\right)\sin\theta$$

or substituting from equation (2-8),

$$\alpha = \frac{\partial v}{\partial x}\cos^2\theta + \left(\frac{\partial v}{\partial y} - \frac{\partial u}{\partial x}\right)\sin\theta\cos\theta - \frac{\partial u}{\partial y}\sin^2\theta \qquad (2\text{-}11)$$

Now, the rotation of the line segment *PT*, which is at right angles to segment *PQ*, is given by equation (2-11) with θ replaced by $\theta + 90°$. This results in:

$$\beta = \frac{\partial v}{\partial x}\sin^2\theta - \left(\frac{\partial v}{\partial y} - \frac{\partial u}{\partial x}\right)\sin\theta\cos\theta - \frac{\partial u}{\partial y}\cos^2\theta \qquad (2\text{-}12)$$

The shear strain γ_{12} which is referred to the 1-2 axes, is the change in the angle between the line segments *PQ* and *PT*. Therefore,

$$\gamma_{12} = \alpha - \beta = \left(\frac{\partial v}{\partial y} - \frac{\partial u}{\partial x}\right) 2 \sin\theta \cos\theta + \left(\frac{\partial v}{\partial x} + \frac{\partial u}{\partial y}\right)(\cos^2\theta - \sin^2\theta)$$

or

$$\frac{1}{2}\gamma_{12} = -\varepsilon_x \sin\theta\cos\theta + \varepsilon_y \sin\theta\cos\theta + \frac{\gamma_{xy}}{2}(\cos^2\theta - \sin^2\theta) \qquad (2\text{-}13)$$

It is now convenient to write the stress and strain transformation equations in matrix form (see Appendix A). The stress transformation equations, equations (2-1), (2-2), and (2-3), in matrix form are:

$$\begin{bmatrix} \sigma_1 \\ \sigma_2 \\ \tau_{12} \end{bmatrix} = \begin{bmatrix} \cos^2\theta & \sin^2\theta & 2\sin\theta\cos\theta \\ \sin^2\theta & \cos^2\theta & -2\sin\theta\cos\theta \\ -\sin\theta\cos\theta & \sin\theta\cos\theta & (\cos^2\theta - \sin^2\theta) \end{bmatrix} \begin{bmatrix} \sigma_x \\ \sigma_y \\ \tau_{xy} \end{bmatrix} \qquad (2\text{-}14)$$

The strain transformation equations, equations (2-9), (2-10), and (2-13), in matrix form can be expressed as:

$$\begin{bmatrix} \varepsilon_1 \\ \varepsilon_2 \\ \frac{1}{2}\gamma_{12} \end{bmatrix} = \begin{bmatrix} \cos^2\theta & \sin^2\theta & 2\sin\theta\cos\theta \\ \sin^2\theta & \cos^2\theta & -2\sin\theta\cos\theta \\ -\sin\theta\cos\theta & \sin\theta\cos\theta & (\cos^2\theta - \sin^2\theta) \end{bmatrix} \begin{bmatrix} \varepsilon_x \\ \varepsilon_y \\ \frac{1}{2}\gamma_{xy} \end{bmatrix} \qquad (2\text{-}15)$$

Equations (2-14) and (2-15) are the transformation relations for the stress or the strain tensor in two dimensional space. They represent nothing more than the Mohr's circle relationships expressed as a matrix equation. They are the necessary relationship required to transform any two dimensional stress or strain state from one set of coordinates to another set of coordinates. They are completely independent of the material under consideration; however, the material must be a continuum. If the transformation matrix utilized in Equation (2-14) and (2-15) is given the Symbol (T), these relations may be expressed as:

$$\begin{bmatrix} \sigma_1 \\ \sigma_2 \\ \tau_{12} \end{bmatrix} = [T] \begin{bmatrix} \sigma_x \\ \sigma_y \\ \tau_{xy} \end{bmatrix}$$

$$\begin{bmatrix} \varepsilon_1 \\ \varepsilon_2 \\ \frac{1}{2}\gamma_{12} \end{bmatrix} = [T] \begin{bmatrix} \varepsilon_x \\ \varepsilon_y \\ \frac{1}{2}\gamma_{xy} \end{bmatrix}$$

Since the components of the stress and strain matrices given by equation (2-14) and (2-15) are transformed by the T matrix, they represent tensors of the second rank in two dimensional space. A discussion of the fundamentals of tensors is given in Appendix A.

2.2 GENERALIZED HOOKE'S LAW [2-1] [2-2]

In elementary strength of materials, it was brought out that Hooke's law is the functional relationship between stress and strain. A laminated composite can have a Hooke's Law relationship which can be classified anywhere from homogeneous isotropic to heterogeneous anisotropic. For this reason, a brief discussion is in order of the elastic symmetries and number of independent elastic constants which describe the behavior of isotropic, orthotropic, and anisotropic materials.

For a homogeneous isotropic material in a one dimensional stress state, the Hooke's law relationship is

$$\sigma = E\,\varepsilon$$

The proportionality constant, E, is Young's modulus or the modulus of elasticity. For the homogeneous isotropic material in a one dimensional stress state, only one elastic constant, E, needs to be specified. For two or three dimensional stress states, an additional elastic constant must be utilized to obtain the stress-strain relations. A second elastic constant, Poisson's ratio, must be determined in order to specify the stress-strain or constitutive relationships for a two or three dimensional stress state in a homogeneous isotropic material. For example, a plane stress state is governed by the following relations:

$$\begin{aligned} \sigma_1 &= (\varepsilon_1 + \nu\varepsilon_2)\,\frac{E}{1-\nu^2} \\ \sigma_2 &= (\varepsilon_2 + \nu\varepsilon_1)\,\frac{E}{1-\nu^2} \\ \tau_{12} &= (\gamma_{12})\,G \end{aligned} \qquad (2\text{-}16)$$

Two *independent* material elastic constants appear in equation (2-16). The third elastic constant, shear modulus, G, is a function of the other two elastic constants, E and ν. This relationship is given by

$$G = E/2(1 + \nu)$$

Therefore, for ISOTROPIC materials, *two* independent elastic constants are necessary to write the Hooke's law relationships for two or three dimensional stress states.

The Hooke's Law relationships for orthotropic and anisotropic materials in two and three dimensional stress states are more involved than the relationships for isotropic materials. The most general form for the Hooke's Law relationships is for anisotropic materials in a three dimensional stress state. For this case, it may be shown that there are 21 independent elastic constants required. If a homogeneous anisotropic material is subjected to a two dimensional stress state (i.e., plane stress or plane strain), then the number of Hooke's Law constants required to determine the inplane stress-strain response is six.

As was pointed out in Chapter 1, an orthotropic material is one which has three planes of material property symmetry. This makes it a more specialized case than the anisotropic material. The number of Hooke's Law or elastic constants required to specify stress-strain relations for a three dimensional stress or strain state in a homogeneous orthotropic medium is nine. If a homogeneous orthotropic medium is subjected to a two dimensional stress or strain state, only four independent elastic constants are required. The orthotropic material subjected to a plane stress state is most pertinent in the study of laminated composites. It will be shown in the next section that the individual plies or laminae of a laminated composite can be assumed to behave as a homogeneous orthotropic material subjected to a plane stress state.

In order to provide a background for the Hooke's Law relations for the orthotropic lamina, which will be presented in the next section, the generalized Hooke's law for the anisotropic material will be presented and subsequently simplified for orthotropic materials. This should clarify any questions which might arise as to the origin of the subscripts on stress, strain, or the Hooke's law constants.

For a three dimensional stress state, such as illustrated in Figure 2-4, the generalized Hooke's Law relation for an anisotropic material is given by:

$$\begin{bmatrix} e_{11} \\ e_{22} \\ e_{33} \\ e_{23} \\ e_{31} \\ e_{12} \\ e_{32} \\ e_{13} \\ e_{21} \end{bmatrix} = \begin{bmatrix} S_{1111} & S_{1122} & S_{1133} & S_{1123} & S_{1131} & S_{1112} & S_{1132} & S_{1113} & S_{1121} \\ S_{2211} & S_{2222} & S_{2233} & S_{2223} & S_{2231} & S_{2212} & S_{2232} & S_{2213} & S_{2221} \\ \text{---} & \text{---} & \text{---} & \text{---} & \text{---} & \text{---} & \text{---} & \text{---} & \text{---} \\ \text{---} & \text{---} & \text{---} & \text{---} & \text{---} & \text{---} & \text{---} & \text{---} & \text{---} \\ \text{---} & \text{---} & \text{---} & \text{---} & \text{---} & \text{---} & \text{---} & \text{---} & \text{---} \\ \text{---} & \text{---} & \text{---} & \text{---} & \text{---} & \text{---} & \text{---} & \text{---} & \text{---} \\ \text{---} & \text{---} & \text{---} & \text{---} & \text{---} & \text{---} & \text{---} & \text{---} & \text{---} \\ \text{---} & \text{---} & \text{---} & \text{---} & \text{---} & \text{---} & \text{---} & \text{---} & \text{---} \\ S_{2111} & S_{2122} & S_{2133} & S_{2123} & S_{2131} & S_{2112} & S_{2132} & S_{2113} & S_{2121} \end{bmatrix} \begin{bmatrix} \sigma_{11} \\ \sigma_{22} \\ \sigma_{33} \\ \sigma_{23} \\ \sigma_{31} \\ \sigma_{12} \\ \sigma_{32} \\ \sigma_{13} \\ \sigma_{21} \end{bmatrix} \tag{2-17}$$

The $[S]$ matrix is the compliance matrix which gives the strain-stress relations for the material. The inverse of the compliance matrix is the stiffness matrix, and it gives the stress-strain relations. There are 81 elastic constants present in the compliance matrix shown above. It can be proven that the stress, strain, and compliance and stiffness matrices must be symmetrical [2-3]. In other words,

$$\sigma_{12} = \sigma_{21}$$
$$\sigma_{23} = \sigma_{32}$$
$$\sigma_{13} = \sigma_{31}$$

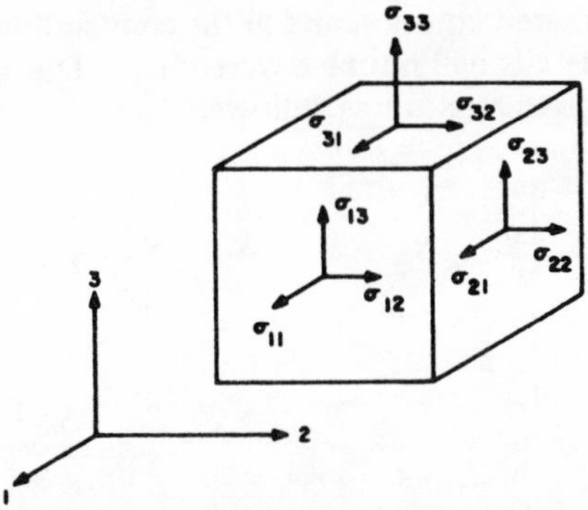

Figure 2-4. Three dimensional state of stress.

and

$$e_{12} = e_{21}$$
$$e_{23} = e_{32} \qquad (2\text{-}18)$$
$$e_{13} = e_{31}$$

and

$$S_{1122} = S_{2211}$$
$$S_{1121} = S_{2111}$$
$$etc.$$

Therefore, with these symmetries, the 81 elastic constants will be reduced to 21 independent elastic constants. This will allow the relations expressed by equation (2-17) to be reduced in size since there are only six independent stresses and strains and 36 components of the S matrix of which only 21 are independent. At this point, it is convenient to change equation (2-17) to contracted notation which will merely shorten the necessary number of subscripts required. The stresses and strains are second order tensors (see Appendix A), and, consequently, the normal notation requires two subscripts. However, it is possible to make the following notation changes

$$\begin{aligned} \sigma_{11} &= \sigma_1 & e_{11} &= \varepsilon_1 \\ \sigma_{22} &= \sigma_2 & e_{22} &= \varepsilon_2 \\ \sigma_{33} &= \sigma_3 & e_{33} &= \varepsilon_3 \\ \sigma_{23} &= \sigma_4 = \tau_{23} & 2e_{23} &= \varepsilon_4 = \gamma_{23} \\ \sigma_{13} &= \sigma_5 = \tau_{13} & 2e_{13} &= \varepsilon_5 = \gamma_{13} \\ \sigma_{12} &= \sigma_6 = \tau_{12} & 2e_{12} &= \varepsilon_6 = \gamma_{12} \end{aligned} \qquad (2\text{-}19)$$

The symmetries expressed by equation (2-18) have been included. The compliances are components of a fourth order tensor; and hence, four subscripts are normally required. The relationships between the stiffnesses and compliances in the contracted notation and in the conventional notation are given by Tsai [2-3]. The details will not be covered here. The end result of the conversion is a constitutive relation as follows:

$$\begin{bmatrix} \varepsilon_1 \\ \varepsilon_2 \\ \varepsilon_3 \\ \varepsilon_4 \\ \varepsilon_5 \\ \varepsilon_6 \end{bmatrix} = \begin{bmatrix} \varepsilon_1 \\ \varepsilon_2 \\ \varepsilon_3 \\ \gamma_{23} \\ \gamma_{13} \\ \gamma_{12} \end{bmatrix} = \begin{bmatrix} S_{11} & S_{12} & S_{13} & S_{14} & S_{15} & S_{16} \\ S_{12} & S_{22} & S_{23} & S_{24} & S_{25} & S_{26} \\ — & — & — & — & — & — \\ — & — & — & — & — & — \\ — & — & — & — & — & — \\ S_{16} & S_{26} & S_{36} & S_{46} & S_{56} & S_{66} \end{bmatrix} \begin{bmatrix} \sigma_1 \\ \sigma_2 \\ \sigma_3 \\ \tau_{23} \\ \tau_{13} \\ \tau_{12} \end{bmatrix} \qquad (2\text{-}20)$$

Use has been made of the symmetry conditions presented in the last of equations (2-18). Now, the strain-stress relations presented in equation (2-20) are for an anisotropic material in a three dimensional stress state. The specially orthotropic material in a three dimensional stress state has the following constitutive relationship:

$$\begin{bmatrix} \varepsilon_1 \\ \varepsilon_2 \\ \varepsilon_3 \\ \gamma_{23} \\ \gamma_{13} \\ \gamma_{12} \end{bmatrix} = \begin{bmatrix} S_{11} & S_{12} & S_{13} & 0 & 0 & 0 \\ S_{12} & S_{22} & S_{23} & 0 & 0 & 0 \\ S_{13} & S_{23} & S_{33} & 0 & 0 & 0 \\ 0 & 0 & 0 & S_{44} & 0 & 0 \\ 0 & 0 & 0 & 0 & S_{55} & 0 \\ 0 & 0 & 0 & 0 & 0 & S_{66} \end{bmatrix} \begin{matrix} \sigma_1 \\ \sigma_2 \\ \sigma_3 \\ \tau_{23} \\ \tau_{13} \\ \tau_{12} \end{matrix} \qquad (2\text{-}21)$$

Note that there are nine independent elastic constants in the compliance matrix. In order to specialize the relations for a two dimensional or plane stress state, the following stresses are assumed zero:

$$\begin{aligned} \sigma_3 &= 0 \\ \tau_{23} &= 0 \\ \tau_{13} &= 0 \end{aligned} \qquad (2\text{-}22)$$

When these terms are set to zero in equation (2-21), it is evident that

$$\begin{aligned} \gamma_{23} &= 0 \\ \gamma_{13} &= 0 \end{aligned} \qquad (2\text{-}23)$$

and

$$\varepsilon_3 = S_{13}\sigma_1 + S_{23}\sigma_2$$

Thus, ε_3 cannot be an independent component since it is a function of the other normal strains, and it may be dropped from the relations. By substituting equations (2-22) and (2-23) into (2-21), the following is obtained:

$$\begin{bmatrix} \varepsilon_1 \\ \varepsilon_2 \\ 0 \\ 0 \\ 0 \\ \gamma_{12} \end{bmatrix} = \begin{bmatrix} S_{11} & S_{12} & S{\uparrow}_{13} & 0 & 0 & 0 \\ S_{12} & S_{22} & S{\uparrow}_{23} & 0 & 0 & 0 \\ S{\uparrow}_{13} & S{\uparrow}_{23} & S{\uparrow}_{33} & 0 & 0 & 0 \\ 0 & 0 & 0 & S{\uparrow}_{44} & 0 & 0 \\ 0 & 0 & 0 & 0 & S{\uparrow}_{55} & 0 \\ 0 & 0 & 0 & 0 & 0 & S_{66} \end{bmatrix} \begin{bmatrix} \sigma_1 \\ \sigma_2 \\ 0 \\ 0 \\ 0 \\ \tau_{12} \end{bmatrix} \qquad (2\text{-}24)$$

There is no point in carrying the extra zeros in the stress, strain, and compliance matrix along; therefore, equation (2-24) can be written in a more compact form:

$$\begin{bmatrix} \varepsilon_1 \\ \varepsilon_2 \\ \gamma_{12} \end{bmatrix} = \begin{bmatrix} S_{11} & S_{12} & 0 \\ S_{12} & S_{22} & 0 \\ 0 & 0 & S_{66} \end{bmatrix} \begin{bmatrix} \sigma_1 \\ \sigma_2 \\ \tau_{12} \end{bmatrix} \qquad (2\text{-}25)$$

Equation (2-25) is the constitutive relationship for the specially orthotropic material in a plane stress state. The presentation which preceded this equation should make it clear to the origin of the S_{66} term and the apparent inconsistencies in the subscripts of the stresses and strains. An excellent presentation of the mechanics of composite materials approached from tensor analysis is given by Tsai in Reference [2-3].

2.3 LAMINA CONSTITUTIVE RELATIONSHIP

In the elementary theory of beams and plates, certain assumptions are made regarding the stress distribution. Certain stresses are assumed to predominate, and other stresses are neglected. In the theory of bending of beams and plates, the normal stress, σ_3, which is perpendicular to the beam or plate midplane, is assumed to be negligible in comparison to the normal stresses, σ_1 or σ_2. In other words, due to the geometry of the beam or plate, the magnitudes that σ_3 can assume are several orders of magnitude less than the values of σ_1 and σ_2 which are induced by the bending. Also, the assumption is made that any line perpendicular to the beam or plate midplane before deformation remains perpendicular to the midplane after deformation and it suffers neither extension nor contraction. As a result, the shear strains, γ_{13} and γ_{23}, and the normal strain, ε_3, are zero. As a consequence, the shear stresses, τ_{13} and τ_{23}, can also be neglected. These assumptions, which are made on the mode of deformation of the beam or plate, are termed the Kirchoff Hypothesis. For thin plates, the hypothesis results in the existence of a plane stress state such as shown in Figure 2-5. Thus, one pertinent assumption in establishing the constitutive or stress-strain relationships for the laminae of a laminated composite is that the laminae, when in the composite, are in a plane stress state. This is not to say that the interlaminar shear stresses, τ_{13} and τ_{23}, will not be present between laminae once they are placed in the composite. However, these stresses may be neglected in establishing the laminae constitutive relations on which the laminate stress-strain relations

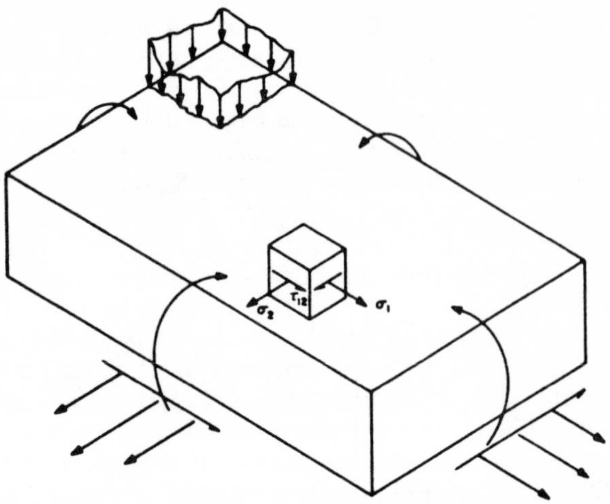

Figure 2-5. State of plane stress.

will be later formulated. These interlaminar shear stresses can be determined from equilibrium considerations in a manner similar to that used for determining the *VQ/I* shear stresses in beams.

It was stated above that the lamina could be considered to be in a plane stress state. For a HOMOGENEOUS ISOTROPIC material in a plane stress state, the stress-strain or Hooke's law relationships for the K^{TH} lamina are:

$$\sigma_1 = (\varepsilon_1 + \nu\varepsilon_2)\frac{E}{1 - \nu^2}$$

$$\sigma_2 = (\varepsilon_2 + \nu\varepsilon_1)\frac{E}{1 - \nu^2}$$

$$\tau_{12} = (\gamma_{12})\frac{E}{2(1 + \nu)}$$

or, in matrix form, the equations are:

$$\begin{bmatrix} \sigma_1 \\ \sigma_2 \\ \tau_{12} \end{bmatrix}_k = \begin{bmatrix} Q_{11} & Q_{12} & 0 \\ Q_{12} & Q_{22} & 0 \\ 0 & 0 & Q_{66} \end{bmatrix}_k \begin{bmatrix} \varepsilon_1 \\ \varepsilon_2 \\ \gamma_{12} \end{bmatrix}_k \tag{2-26}$$

where,

$$Q_{11} = Q_{22} = E/(1 - \nu^2)$$
$$Q_{12} = \nu E/(1 - \nu^2)$$
$$Q_{66} = E/2(1 + \nu) = G$$

Thus, equation (2-26) is the constitutive or stress-strain relationship for HOMOGENEOUS ISOTROPIC laminae of a laminated composite. It should be emphasized that in the relations described by equation (2-26) there are only two independent material elastic constants, E and ν. For isotropic materials, the third elastic constant, shear modulus, G, is a function of the other two elastic constants, E and ν. The relationship is:

$$G = E/2(1 + \nu)$$

Therefore, for isotropic materials, two elastic constants are necessary to write the Hooke's law or constitutive relations for plane stress. For isotropic materials only, two elastic constants are sufficient to express the constitutive relations for three dimensional stress states also. The isotropic condition also requires that the material has the same properties in tension and compression.

For filamentary laminae, both unidirectional and woven, the constitutive relations may be established by assuming the laminae are homogeneous orthotropic materials and are also in a plane stress state. The geometry of unidirectional and woven filamentary lamina, Figure 2-6, shows that there are three planes of material property symmetry; therefore, as brought out in the previous section, the laminae are orthotropic. The macro mechanics approach to the problem ignores the fiber-resin geometry and interactions and assumes the laminae is a homogeneous medium. Thus, the Hooke's law rela-

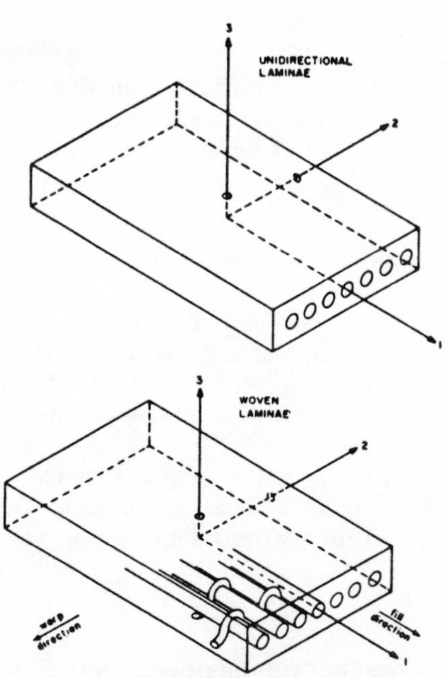

Figure 2-6. *Lamina principal axis.*

tionship for a homogeneous material which has orthotropic material properties and is in a plane stress state may be written as:

$$\begin{aligned}\sigma_1 &= Q_{11}\varepsilon_1 + Q_{12}\varepsilon_2\\ \sigma_2 &= Q_{12}\varepsilon_1 + Q_{22}\varepsilon_2\\ \tau_{12} &= Q_{66}\gamma_{12}\end{aligned}$$

or, in matrix form

$$\begin{bmatrix}\sigma_1\\ \sigma_2\\ \tau_{12}\end{bmatrix}_k = \begin{bmatrix}Q_{11} & Q_{12} & 0\\ Q_{12} & Q_{22} & 0\\ 0 & 0 & 2Q_{66}\end{bmatrix}_k \begin{bmatrix}\varepsilon_1\\ \varepsilon_2\\ \frac{1}{2}\gamma_{12}\end{bmatrix}_k \tag{2-27}$$

where, the components of the Hooke's law or STIFFNESS matrix, Q, are:

$$\begin{aligned}Q_{11} &= E_{11}/(1 - \nu_{12}\nu_{21})\\ Q_{22} &= E_{22}/(1 - \nu_{12}\nu_{21})\\ Q_{12} &= \nu_{21}E_{11}/(1 - \nu_{12}\nu_{21}) = \nu_{12}E_{22}/(1 - \nu_{12}\nu_{21})\\ Q_{66} &= G_{12}\\ Q_{16} &= Q_{26} = 0\end{aligned} \tag{2-28}$$

As expected, there are four independent elastic constants: the Young's moduli in the one and two directions (see Figure 2-6), E_{11} and E_{22}; the shear modulus, G_{12}; the major Poisson's ratio, ν_{12}. The fifth elastic constant, ν_{12}, is a function of the other constants, and it may be determined from the fact that the Hooke's law matrix, Q, must be symmetrical about the main diagonal (reciprocality relation):

$$\nu_{12}E_{11} = \nu_{12}E_{22}$$

Thus, it has been shown that in order to obtain the required Hooke's law constants for a thin orthotropic lamina in a plane stress state, four independent elastic constants must be determined. Therefore, the orthotropic material necessitates the determination of two more elastic constants than is necessary for isotropic materials.

When the lamina stiffness matrix, Q, is inverted, the lamina COMPLIANCE matrix, S, is obtained. That is, the compliance matrix is the inverse (see Appendix A) of the stiffness matrix:

$$[S] = [Q]^{-1} \tag{2-29}$$

Thus, the strain-stress relations for the specially orthotropic lamina are given by:

$$\begin{bmatrix} \varepsilon_1 \\ \varepsilon_2 \\ \gamma_{12} \end{bmatrix}_k = \begin{bmatrix} S_{11} & S_{12} & 0 \\ S_{12} & S_{22} & 0 \\ 0 & 0 & S_{66} \end{bmatrix}_k \begin{bmatrix} \sigma_1 \\ \sigma_2 \\ \tau_{12} \end{bmatrix}_k \tag{2-30}$$

The components of the lamina compliance matrix, S, are given by:

$$\begin{aligned} S_{11} &= 1/E_{11} \\ S_{22} &= 1/E_{22} \\ S_{12} &= -\nu_{12}/E_{11} = -\nu_{21}/E_{22} \\ S_{66} &= 1/G_{12} \\ S_{16} &= S_{26} = 0 \end{aligned} \tag{2-31}$$

As can be seen from equations (2-31), the four elastic constants which characterize the lamina are present. This completes the equations necessary to supply the constitutive relationships for the SPECIALLY ORTHOTROPIC lamina. The term specially orthotropic is utilized to distinguish between the constitutive relationships which are referred to the lamina principal axes (1,2) and those which are referred to an arbitrary set of axes, (x,y).

Normally, the lamina principal axes (1,2) are not coincident with the reference axes for the laminate, (x,y). When this occurs, the constitutive relations for each individual lamina must be transformed to the laminate reference axes in order to determine the laminate constitutive relationship. As was shown earlier, the following transforms may be applied to the stress and strain in the lamina since they are second order tensors:

$$\begin{bmatrix} \sigma_1 \\ \sigma_2 \\ \tau_{12} \end{bmatrix}_k = [T] \begin{bmatrix} \sigma_x \\ \sigma_y \\ \tau_{xy} \end{bmatrix}_k$$

$$\begin{bmatrix} \varepsilon_1 \\ \varepsilon_2 \\ \dfrac{1}{2}\gamma_{12} \end{bmatrix}_k = [T] \begin{bmatrix} \varepsilon_x \\ \varepsilon_y \\ \dfrac{1}{2}\gamma_{xy} \end{bmatrix}_k$$

Or, if the equations are inverted (see Appendix A)

$$[T]\,[T]^{-1} \begin{bmatrix} \sigma_x \\ \sigma_y \\ \tau_{xy} \end{bmatrix}_k = \begin{bmatrix} \sigma_x \\ \sigma_y \\ \tau_{xy} \end{bmatrix}_k = [T]^{-1} \begin{bmatrix} \sigma_1 \\ \sigma_2 \\ \tau_{12} \end{bmatrix}_k$$

$$[T]\,[T]^{-1} \begin{bmatrix} \varepsilon_x \\ \varepsilon_y \\ \dfrac{1}{2}\gamma_{xy} \end{bmatrix}_k = \begin{bmatrix} \varepsilon_x \\ \varepsilon_y \\ \dfrac{1}{2}\gamma_{xy} \end{bmatrix}_k = [T]^{-1} \begin{bmatrix} \varepsilon_1 \\ \varepsilon_2 \\ \dfrac{1}{2}\gamma_{12} \end{bmatrix}_k \tag{2-32}$$

The transformation matrix, T, is given by equation (2-15), or

$$[T] = \begin{bmatrix} m^2 & n^2 & 2mn \\ n^2 & m^2 & -2mn \\ -mn & mn & m^2-n^2 \end{bmatrix}$$

$$m = \cos\theta$$
$$n = \sin\theta$$

Note that the inverse of the T matrix may be obtained by substituting for the positive angle θ, as defined in Figure 2-7, a negative angle θ. By utilizing equations (2-15), (2-27), and (2-32), the lamina stress-strain relations transformed to the laminate reference axes can be shown to be:

$$\begin{bmatrix} \sigma_x \\ \sigma_y \\ \tau_{xy} \end{bmatrix}_k = [T]^{-1}[Q]^k\,[T] \begin{bmatrix} \varepsilon_x \\ \varepsilon_y \\ \frac{1}{2}\gamma_{xy} \end{bmatrix}_k \qquad (2\text{-}33)$$

The matrix multiplications, $[T]^{-1}\,[Q]\,[T]$, which are shown in equation (2-33) can be accomplished as follows:

$$[T]^{-1} = [T(-\theta)] = \begin{bmatrix} m^2 & n^2 & -2mn \\ n^2 & m^2 & 2mn \\ mn & -mn & m^2-n^2 \end{bmatrix}$$

therefore, the first matrix operation becomes (see Appendix A)

$$[T]^{-1}[Q] = \begin{bmatrix} m^2 & n^2 & -2mn \\ n^2 & m^2 & 2mn \\ mn & -mn & m^2-n^2 \end{bmatrix} \begin{bmatrix} Q_{11} & Q_{12} & 0 \\ Q_{12} & Q_{22} & 0 \\ 0 & 0 & 2Q_{66} \end{bmatrix}$$

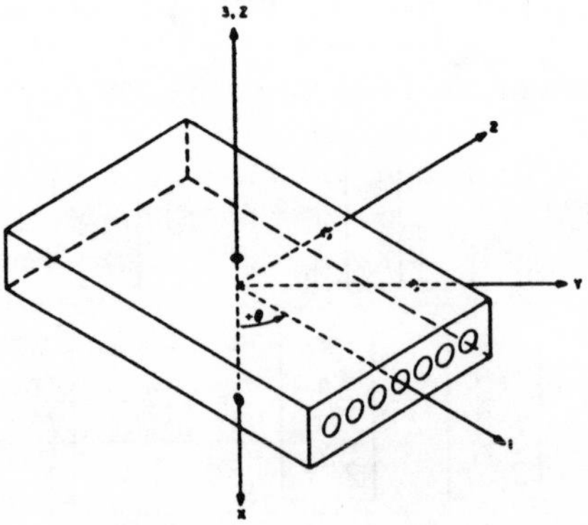

Figure 2-7. Lamina axis rotation.

This multiplication operation results in

$$\begin{bmatrix} (m^2 Q_{11} + n^2 Q_{12}) & (m^2 Q_{12} + n^2 Q_{22}) 2(-2mnQ_{66}) \\ (n^2 Q_{11} + m^2 Q_{12}) & (n^2 Q_{12} + m^2 Q_{22}) 2(2mnQ_{66}) \\ (mn Q_{11} - mn Q_{12}) & (nm Q_{12} - mn Q_{22}) 2(m^2 - n^2) Q_{66} \end{bmatrix}$$

The above matrix is now post multiplied by the *[T]* matrix. If the *1/2* which appeared in the strain matrix is multiplied by the above result, the following is obtained:

$$\begin{bmatrix} \sigma_x \\ \sigma_y \\ \tau_{xy} \end{bmatrix}^k = \begin{bmatrix} \bar{Q}_{12} & \bar{Q}_{12} & \bar{Q}_{16} \\ \bar{Q}_{12} & \bar{Q}_{22} & \bar{Q}_{26} \\ \bar{Q}_{16} & \bar{Q}_{26} & \bar{Q}_{66} \end{bmatrix}^k \begin{bmatrix} \varepsilon_x \\ \varepsilon_y \\ \gamma_{xy} \end{bmatrix}^k \tag{2-34}$$

where the components of the lamina stiffness matrix, $\bar{Q}$, which is now referred to an arbitrary set of axes, are given by:

$$\begin{aligned} \bar{Q}_{11} &= Q_{11} \cos^4\theta + 2(Q_{12} + Q_{66}) \sin^2\theta \cos^2\theta + Q_{22} \sin^4\theta \\ \bar{Q}_{22} &= Q_{11} \sin^4\theta + 2(Q_{12} + 2Q_{66}) \sin^2\theta \cos^2\theta + Q_{22} \cos^4\theta \\ \bar{Q}_{12} &= (Q_{11} + Q_{22} - 4Q_{66}) \sin^2\theta \cos^2\theta + Q_{12} (\sin^4\theta + \cos^4\theta) \\ \bar{Q}_{66} &= (Q_{11} + Q_{22} - 2Q_{12} - 2Q_{66}) \sin^2\theta \cos^2\theta + Q_{66} (\sin^4\theta + \cos^4\theta) \\ \bar{Q}_{16} &= (Q_{11} - Q_{12} - 2Q_{66}) \sin\theta\cos^3\theta + (Q_{12} - Q_{22} + 2Q_{66}) \sin^3\theta\cos\theta \\ \bar{Q}_{26} &= (Q_{11} - Q_{12} - 2Q_{66}) \sin^3\theta\cos\theta + (Q_{12} - Q_{22} + 2Q_{66}) \sin\theta\cos^3\theta \end{aligned} \tag{2-35}$$

The $\bar{Q}$ matrix is now fully populated, and it would seem that there are now six elastic constants which govern the behavior of the lamina. However, $\bar{Q}_{16}$ and $\bar{Q}_{26}$ are merely linear combinations of the four basic elastic constants and are not independent. In the transformed coordinate system, the lamina is said to be "generally" orthotropic, and the $\bar{Q}$ matrix is similar in appearance to the Q matrix for a fully anisotropic lamina $Q \neq 0$, $Q_{26} \neq 0$. Therefore, equation (2-34) is said to be the constitutive equation for a "generally" orthotropic lamina. Equation (2-27) is normally referred to as the constitutive equation for a "specially" orthotropic lamina since $Q_{16} = Q_{26} = 0$. It should be noted that the rotations (T matrix) are with respect to the plate axes rather than the material axes. This convention is consistent with stress analysis procedures in which one knows the state of stress with respect to the plate axes and wishes to determine it with respect to the fiber direction. Materials investigators, however, usually know properties with respect to the material (orthotropic) axes and wish to determine them with respect to the plate axes. If the second convention is used, then T becomes T^{-1}, and a negative sign appears in front of the expressions for $\bar{Q}_{16}$ and $\bar{Q}_{26}$ in equation (2-35). Thus, it is imperative that the reference axes are clearly defined (e.g., see Pagano and Chou, *Journal of Composite Materials,* January, 1969).

A convenient form for the transformed lamina stiffnesses has been given by Tsai and Pagano [2–4]:

$$\begin{aligned}
\bar{Q}_{11} &= U_1 + U_2 \cos(2\theta) + U_3 \cos(4\theta) \\
\bar{Q}_{22} &= U_1 - U_2 \cos(2\theta) + U_3 \cos(4\theta) \\
\bar{Q}_{12} &= U_4 - U_3 \cos(4\theta) \\
\bar{Q}_{66} &= U_5 - U_3 \cos(2\theta) \\
\bar{Q}_{16} &= +\frac{1}{2} U_2 \sin(2\theta) + U_3 \sin(4\theta) \\
\bar{Q}_{26} &= +\frac{1}{2} U_2 \sin(2\theta) - U_3 \sin(4\theta)
\end{aligned} \tag{2-36}$$

where

$$\begin{aligned}
U_1 &= \frac{1}{8}(3Q_{11} + 3Q_{22} + 2Q_{12} + 4Q_{66}) \\
U_2 &= \frac{1}{2}(Q_{11} - Q_{22}) \\
U_3 &= \frac{1}{8}(Q_{11} + Q_{22} - 2Q_{12} - 4Q_{66}) \\
U_4 &= \frac{1}{8}(Q_{11} + Q_{22} + 6Q_{12} - 4Q_{66}) \\
U_5 &= \frac{1}{8}(Q_{11} + Q_{22} - 2Q_{12} + 4Q_{66})
\end{aligned} \tag{2-37}$$

and $$U_5 = \frac{1}{2}(U_1 - U_4)$$

and, U_1, U_4, and U_5 are invariant to a rotation about the three axis. Note that only two of the invariants, U_1, U_4, and U_5 are independent. That is, these terms remain constant regardless of the angle θ. Thus, examining equation (2-36) reveals that each of the first four terms (independent terms) is composed of a constant term plus terms which change with the angle θ. Therefore, there are certain inherent lamina properties which are only dependent on the material being used and do not change with orientation of the lamina. This "invariant" concept is very useful in design with composite materials.

The next case of interest is the lamina composed of a HOMOGENEOUS ANISOTROPIC material. This type of material has no planes of material property symmetry such as the orthotropic material. It is difficult to portray what the anisotropic lamina would be like such as was done for the orthotropic laminae in Figure 2-6. The anisotropic lamina behaves in a manner very similar to the HOMOGENEOUS GENERALLY ORTHOTROPIC lamina. As was pointed out earlier, an anisotropic material in a plane stress state has six independent elastic constants; therefore, the lamina constitutive relations are given by:

$$\begin{bmatrix} \sigma_1 \\ \sigma_2 \\ \tau_{12} \end{bmatrix} = \begin{bmatrix} Q_{11} & Q_{12} & Q_{16} \\ Q_{12} & Q_{22} & Q_{26} \\ Q_{16} & Q_{26} & Q_{66} \end{bmatrix} \begin{bmatrix} \varepsilon_1 \\ \varepsilon_2 \\ \gamma_{12} \end{bmatrix} \tag{2-38}$$

There is no direct correlation between the six elastic constants given in equation (2-38), and the engineering constants such as there was for the orthotropic material. It is evident that the constitutive equation for the anisotropic lamina has the same general form as the equation for the generally orthotropic lamina; however, the constants Q_{16} and Q_{26} are independent material constants.

As an example of the behavior of the stiffness matrix terms for an orthotropic lamina as the coordinate system is rotated, the following example is presented. The orthotropic lamina has the following elastic constants:

$$\begin{aligned} E_{11} &= 30.0 \times 10^6 \\ E_{22} &= 3.0 \times 10^6 \\ G_{12} &= 1.0 \times 10^6 \\ \nu_{12} &= 0.3 \end{aligned}$$

These constants are representative of a boron/epoxy composite material. The lamina stiffness or Hooke's law matrix is as follows:

$$[Q] = \begin{bmatrix} 30.30 & 0.91 & 0 \\ 0.91 & 3.03 & 0 \\ 0 & 0 & 1.00 \end{bmatrix} 10^6$$

The Q matrix, which is a function of the angle of orientation of the x,y axes, is given by equation (2-35) or (2-36). The terms of the Q matrix taken from equation (2-36) are:

$$\begin{aligned} \bar{Q}_{11} &= (13.35 \times 10^6) + (13.64 \times 10^6) \cos 2\theta + (3.44 \times 10^6) \cos 4\theta \\ \bar{Q}_{22} &= (13.35 \times 10^6) - (13.64 \times 10^6) \cos 2\theta + (3.44 \times 10^6) \cos 4\theta \\ \bar{Q}_{12} &= (4.35 \times 10^6) - (3.44 \times 10^6) \cos 4\theta \\ \bar{Q}_{66} &= (4.48 \times 10^6) - (3.44 \times 10^6) \cos 4\theta \\ \bar{Q}_{16} &= +\frac{1}{2}(13.64 \times 10^6) \sin 2\theta + (3.44 \times 10^6) \sin 4\theta \\ \bar{Q}_{26} &= +\frac{1}{2}(13.64 \times 10^6) \sin 2\theta - (3.44 \times 10^6) \sin 4\theta \end{aligned}$$

and

$$\begin{aligned} U_1 &= 13.35 \times 10^6 \\ U_2 &= 13.64 \times 10^6 \\ U_3 &= 3.44 \times 10^6 \\ U_4 &= 4.35 \times 10^6 \\ U_5 &= 4.48 \times 10^6 \end{aligned}$$

In Figures (2-8) and (2-9), the $\bar{Q}$ terms are plotted as a function of the angle θ. Note that the behavior of the $\bar{Q}_{11}$ and $\bar{Q}_{22}$ terms, the $\bar{Q}_{12}$ and $\bar{Q}_{66}$ terms, and

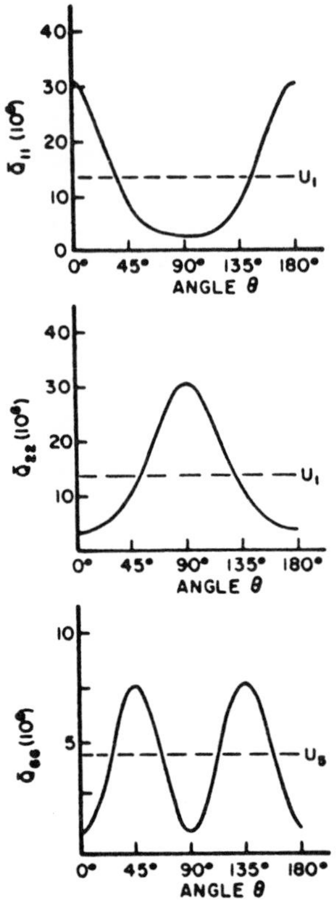

Figure 2-8. *Lamina stiffness versus angle of rotation.*

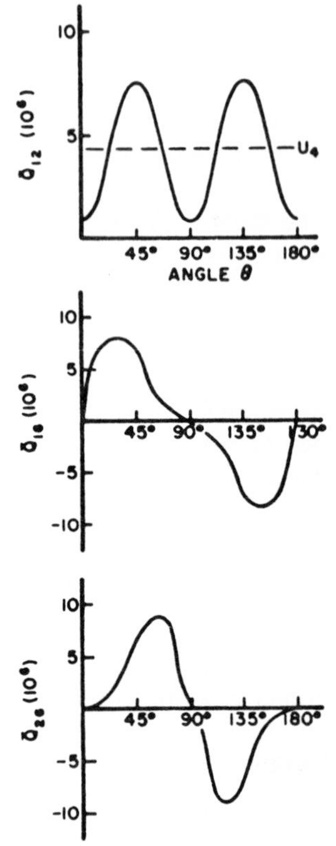

Figure 2-9. *Lamina stiffness versus angle of rotation.*

the $\bar{Q}_{16}$ and $\bar{Q}_{26}$ terms are similar. The invariant quantities, U, are shown in the figures.

2.4 SHEAR COUPLING PHENOMENA

The shear and normal components of stress and strain in Equations (2-26) and (2-30) are uncoupled in the principle material coordinate systems. That is, normal stresses produce only normal strains and shear stresses produce only shear strain. In an arbitrary coordinate system, Equation (2-34), the shear and normal response is coupled through the shear coupling compliance and stiffness terms $\bar{S}_{16}$, $\bar{S}_{26}$, $\bar{Q}_{16}$, and $\bar{Q}_{26}$. As a quantitative measure of this interaction, the shear coupling ratios have been defined.

$$\eta_{xy} = \frac{\gamma_{xy}}{\varepsilon_x} \quad when\ \sigma_y = \tau_{xy} = 0$$

$$\eta_{yx} = \frac{\gamma_{xy}}{\varepsilon_y} \quad when\ \sigma_x = \tau_{xy} = 0 \tag{2-39}$$

The shear coupling ratio can be expressed as a function of the compliance terms

$$\eta_{xy} = \bar{S}_{16}/\bar{S}_{11}$$
$$\eta_{yx} = \bar{S}_{26}/\bar{S}_{22} \tag{2-40}$$

when the fully populated matrix of Equation (2-30) are combined with Equation (2-39). The generic uncoupled and coupled responses are illustrated in Figures 2-10 and 7-1. Figure 2-11 consists of photographs of unidirectional nylon cord reinforced elastomer type ply material at fiber axis orientation of 30°, 60°, and 75° with respect to the principle load direction. For highly anisotropic systems, such as this, the shear coupling rotation will change in direction with increased off-axis orientation (see Example 2-1). The equivalent engineering coefficients for the nylon-elastomer ply system is shown in Figure 2-12. This material system will be tracted throughout this primer as a means of visually illustrating anisotropic phenomena.

Example 2-1

To illustrate the characteristics of an orthotropic lamina, the following problem is outlined. Consider a nylon reinforced elastomeric lamina, the ply of a tyre with the following properties:

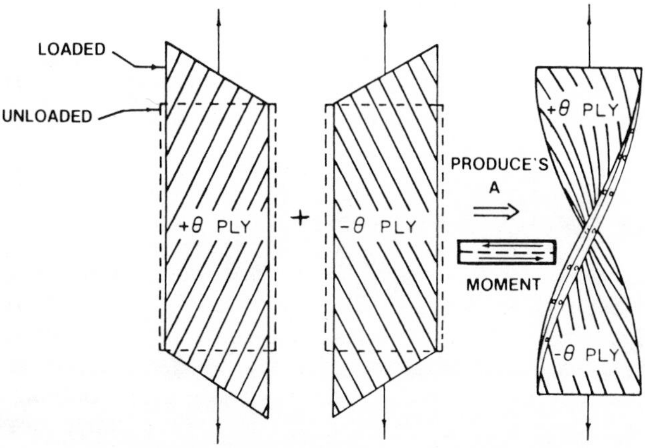

Figure 2-10. *Illustration of shear coupling deformation.*

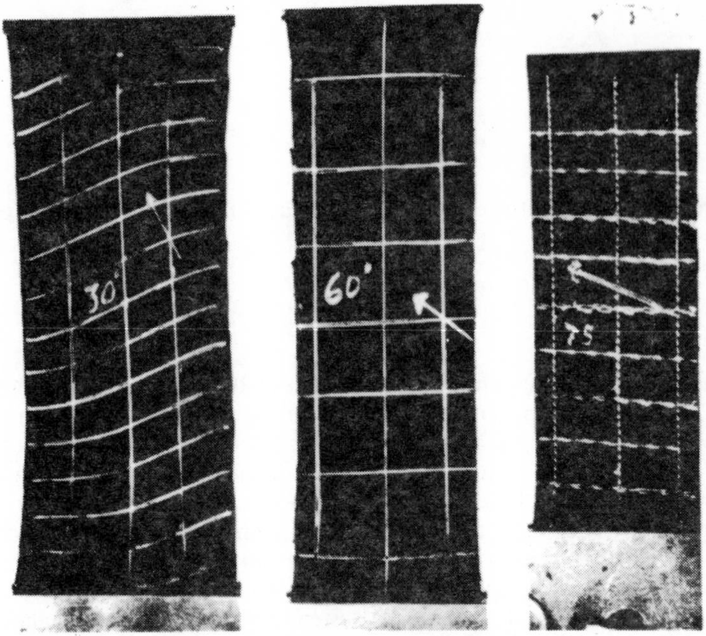

Figure 2-11. Unaxial off-axis test: compliant end condition.

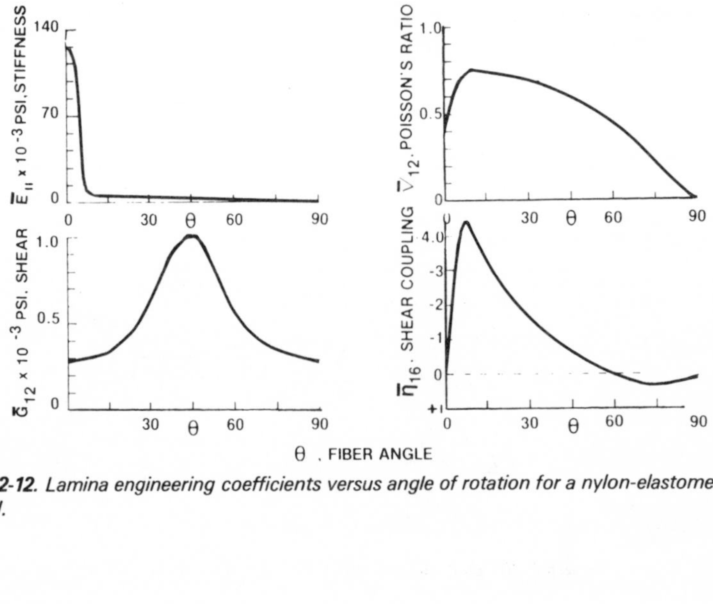

***Figure 2-12.** Lamina engineering coefficients versus angle of rotation for a nylon-elastomeric ply material.*

$$E_{11} = 132{,}000 \ PSI$$
$$E_{22} = 1{,}050 \ PSI$$
$$\nu_{12} = 0.36$$
$$G_{12} = 263 \ PSI$$

Compute the: (a) off-axis engineering moduli for a ply as the fiber axis is rotated with respect to the load direction. Chapter 7, Section 7.2 discusses the characterization of an orthotropic including the general form of Equation (2-20). The off-axis engineering moduli is generally measured as an off-axis compliance.

For a unaxial tension test the conditions are

$$\sigma_1 > 0$$
$$\sigma_2 = \sigma_6 = 0$$

and the response according to Section 7.2 is

$$\varepsilon_1 = S_{11}\sigma_1 = \frac{\sigma_1}{E_{11}}$$
$$\varepsilon_2 = S_{21}\sigma_1 = -\frac{\nu_{21}}{E_{22}}\sigma_1 = -\frac{\nu_{12}}{E_{11}}\sigma_1$$
$$\gamma_{12} = \varepsilon_6 = S_{61}\sigma_1 = \frac{\eta_{61}}{E_1}\sigma_1$$

The symmetry conditions from Equation (2-18) yield

$$S_{12} = S_{21}$$
$$S_{16} = S_{61}$$

with

$$\nu_{12} = \frac{\varepsilon_1}{\varepsilon_2}$$
$$\eta_{16} = \frac{\varepsilon_1}{\gamma_{12}}$$

Following Equations (2-35) through (2-37) we can write:

$$\bar{S}_{11} = U_1 + U_2 \cos 2\theta + U_3 \cos 4\theta$$
$$\bar{S}_{22} = U_1 - U_2 \cos 2\theta + U_3 \cos 4\theta$$
$$\bar{S}_{21} = \bar{S}_{12} = U_4 - U_3 \cos 4\theta$$
$$\bar{S}_{66} = U_5 - 4U_3 \cos 4\theta$$
$$\bar{S}_{61} = \bar{S}_{16} = U_2 \sin 2\theta + 2U_3 \sin 4\theta$$
$$\bar{S}_{62} = \bar{S}_{26} = U_2 \sin 2\theta - 2U_3 \sin 4\theta$$

The definitions of the U's for the S_{11} transformation are

$$U_1 = \frac{1}{8}(3S_{11} + 3S_{22} + 2S_{12} + S_{66})$$

$$U_2 = \frac{1}{2}(S_{11} - S_{22})$$

$$U_3 = \frac{1}{8}(S_{11} + S_{22} - 2S_{12} - S_{66})$$

$$U_4 = \frac{1}{8}(S_{11} + S_{22} + 6S_{12} - S_{66})$$

$$U_5 = \frac{1}{2}(S_{11} + S_{22} - 2S_{12} + S_{66})$$

The difference between the invarients for the compliances versus the moduli, Equations (2-36) and (2-37) can be traced to the use of engineering shear strain. The specific compliances for the problem are:

$$\frac{1}{E_{11}} = S_{11} = (132{,}000 \text{ psi})^{-1} = 7.58 \times 10^{-6} \text{ psi}^{-1}$$

$$\frac{1}{E_{22}} = S_{22} = (1{,}050 \text{ psi})^{-1} = 9.52 \times 10^{-4} \text{ psi}^{-1}$$

$$-\nu_{12}S_{22} = -\frac{\nu_{12}}{E_{11}} = S_{12} = -\left(\frac{0.36}{132{,}000 \text{ psi}}\right) = -2.73 \times 10^{-6} \text{ psi}^{-1}$$

$$\frac{1}{G_{12}} = S_{66} = (263 \text{ psi})^{-1} = 3.8 \times 10^{-3} \text{ psi}^{-1}$$

$$S_{11} = \frac{1}{E_{11}} = 7.58 \times 10^{-6} \text{ psi}^{-1}$$

$$S_{22} = \frac{1}{E_{22}} = 9.52 \times 10^{-6} \text{ psi}^{-1}$$

$$S_{12} = -\frac{\nu_{12}}{E_{11}} = -2.73 \times 10^{-6} \text{ psi}^{-1}$$

$$S_{66} = \frac{1}{G_{12}} = 3.8 \times 10^{-3} \text{ psi}$$

The off-axis compliances are then determined to be:

θ	$\bar{S}_{11}(psi^{-1})$	$\bar{S}_{12}(psi^{-1})$	$\bar{S}_{66}(psi^{-1})$	$\bar{S}_{16}(psi^{-1})$
0	7.58×10^{-6}	-2.73×10^{-6}	3.80×10^{-3}	—
15	2.49×10^{-4}	-1.85×10^{-4}	3.09×10^{-3}	-8.50×10^{-4}
30	7.78×10^{-4}	-5.35×10^{-4}	1.67×10^{-3}	-1.02×10^{-3}
45	1.19×10^{-3}	-7.12×10^{-4}	9.65×10^{-4}	-4.72×10^{-4}
60	1.25×10^{-3}	-5.35×10^{-4}	1.67×10^{-3}	2.05×10^{-4}
75	1.07×10^{-3}	-1.85×10^{-4}	3.09×10^{-3}	3.78×10^{-6}
90	9.52×10^{-4}	-2.73×10^{-6}	3.80×10^{-3}	—

To obtain the effective engineering moduli the $\bar{S}_{11}$ and $\bar{S}_{66}$ quantities are inverted. The other two quantities are computed as follows:

$$\nu_{12} = \bar{E}_{11} \times \bar{S}_{12}$$
$$\eta_{12} = \bar{E}_{11} \times \bar{S}_{16}$$

θ	$\bar{E}_{11}(psi)$	ν_{12}	$G_{12}(psi)$	$\bar{\eta}_{16}$
0	131,963	0.36	263	0.0
15	4,018	0.72	323	-3.4
30	1,286	0.69	597	-1.3
45	839	0.60	1036	-0.86
60	800	0.43	597	+0.16
75	937	0.17	323	+0.35
90	1,050	0.003	263	0.00

The correlation of the moduli $\bar{E}_{11}$ with measured values is shown in Figure 3-6.

2.5 THERMAL AND ENVIRONMENTAL INDUCED DISTORATION OF ORTHOTROPIC LAMINA

A change in temperature or the absorption of fluids or gases from the environmental will result in a dimensional change in a lamina. When an orthotropic material (lamina) is subjected to a change in temperature or absorption induced swelling, the stress-strain relations, Equations (2-16)–(2-34), must be modified to account for the stress free environmentally induced expansional strains; e_i, Equation (2-16) may be rewritten as follows:

$$\varepsilon_i = S_{ij}\,\sigma_j + e_i$$

or (2-41)

$$\sigma_i = C_{ij}\,(\varepsilon_j - e_i)$$

These expressions are known as the Duhammel-Neumann [2-5] form of Hooke's Law. In expanded form Equation (2-40) appears as

$$\begin{aligned} \varepsilon_1 &= S_{11}\,\sigma_1 + S_{12}\,\sigma_2 + e_1 \\ \varepsilon_2 &= S_{12}\,\sigma_1 + S_{22}\,\sigma_2 + e_2 \\ \gamma_{12} &= S_{66}\,\tau_{12} \end{aligned} \qquad (2\text{-}42)$$

When it is desired to determine the stress-strain relation in an arbitrary coordinate system it is necessary to determine the expansional strains, e, in that coordinate system.

$$\begin{aligned} \bar{e}_x &= m^2\,\bar{e}_1 + n^2\,\bar{e}_2 \\ \bar{e}_y &= n^2\,\bar{e}_1 + m^2\,\bar{e}_2 \\ \bar{e}_{xy} &= 2mn\,\bar{e}_1 - 2mn\,\bar{e}_2 \\ m &= \cos\theta \text{ and } n = \sin\theta \end{aligned} \qquad (2\text{-}43)$$

It should be noted that the term e_{xy} vanishes in general only a θ-0 deg., and 90 deg. e_{xy} is a shear coefficient of environmentally induced expansion. When $e_{xy} \neq 0$ a change in environment will produce a shear strain in the x-y coordinate system. The strains e_1 and e_2 $(= e_3)$ characterize the mechanical consequences of temperature variations in service conditions, consequences of molding or curing shrinkage, the effects of swelling agents upon composite structures, etc.

Example 2-2

To illustrate the strain transformation calculations and the induced expansional shear deformation compute the strains defined in Equation (2-43). The material in Example 2-1 was soaked in a solvent resulting in a large volumetric expansion, Figure 2-13. The measured [2-6] expansional strains were:

$$e_1 = 0.011$$
$$e_2 = 0.75$$

The computed strains are

θ	$\bar{e}_1$	$\bar{e}_6 = \bar{e}_{12}$
0	0.011	—
15	0.061	0.185
30	0.220	0.320
45	0.381	0.370
60	0.565	0.320
75	0.700	0.185
90	0.750	—

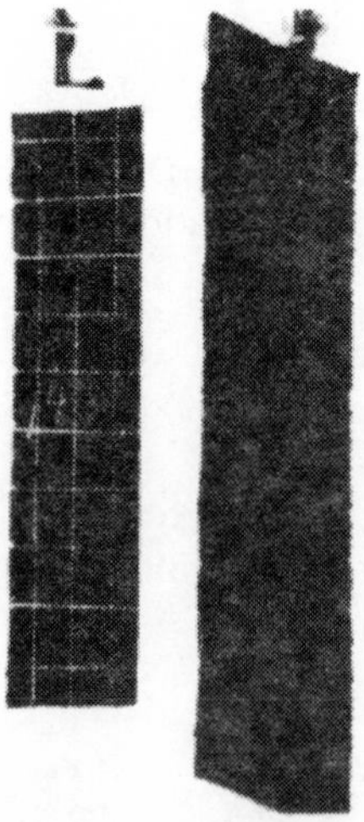

Figure 2-13. *Comparison of a dry and a swollen 45 degree off-axis specimen.*

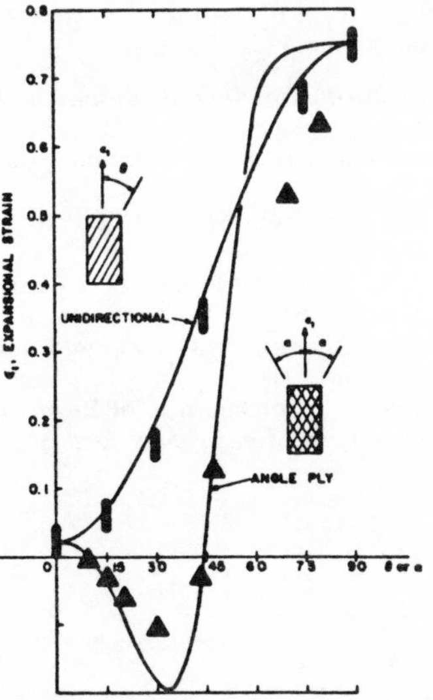

Figure 2-14. *Comparison of theoretical and experimentally observed expansional strains for unidirectional lamina and angle ply laminates of nylon-elastomeric ply material.*

Figure 2-13 consists of a rectangular strip of the material at a fiber orientation of 45 ± to the major axis of the rectangle. The magnitude of the strains can be appreciated through the comparison of the "dry" and "swollen" specimens. The correlation with measurements is shown in Figure 2-14.

The general expression for Hooke's Law is

$$\begin{aligned} \varepsilon_x &= S_{11}\,\sigma_x + S_{12}\,\sigma_y + S_{16}\,\tau_{xy} + e_x \\ \varepsilon_y &= S_{12}\,\sigma_x + S_{22}\,\sigma_y + S_{26}\,\tau_{xy} + e_y \\ \gamma_{xy} &= S_{16}\,\sigma_x + S_{26}\,\sigma_y + S_{66}\,\tau_{xy} + e_{xy} \end{aligned} \tag{2-44}$$

The general expansional strains, e_i, is expressed as

$$\bar{e}_i = \alpha_i\,\Delta T + \beta_i\,C \tag{2-45}$$

Where α_i and β_i are thermal expansion and swelling (moisture, etc.) expansion coefficients, respectively. Temperature is denoted by T and swelling agent concentration (moisture, etc.) by C. The two terms are emphasized because in polymeric composites, wood laminates, tyre construction, etc., volume changes due to absorbed swelling agents (for example, moisture from a humid environment) are equally as important as a volume change due to temperature. Thus, the equivalent of a temperature change or the absorption of a swelling agent (liquid or vapor) is a mechanical strain or stress.

REFERENCES

1. Hearmon, R. F. S. *An Introduction to Applied Anisotropic Elasticity.* Oxford University Press (1961).
2. Lekhnitskii, S. G. *Theory of Elasticity of an Anisotropic Elastic Body.* Holden-Day, Inc. (1963).
3. Tsai, S. W., "Mechanics of Composite Materials," Part II, Technical Report AFML-TR-66-149 (November 1966).
4. Tsai, S. W. and Pagano, N. J., "Invariant Properties of Composite Materials," *Composite Materials Workshop.* Technomic Publishing Company (January 1968).
5. Halpin, J. C., "Introduction to Viscoelasticity," *Composite Materials Workshop.* Technomic Publishing Company (January 1968).
6. Halpin, J. C. and Pagano, N. J., "Consequences of Environmentally Induced Dilation in Solids," in *Proc. of the 6th Annual Meeting, Soc. Eng. Sci.,* New York: Springer-Verlag (1969).

3
Laminated Composites

3.1 INTRODUCTION

THIS CHAPTER IS CONCERNED WITH THE PROPERTIES OF LAMINATED COMposite materials, that is, with the properties of multiple layers of materials which act together as a single layer. In the previous chapter, the behavior of a single orthotropic layer has been considered in detail. Since a laminate is composed of several such layers, the description of the behavior of a single layer (of a lamina) forms the basis or building block with which the behavior of a laminate is described. In this chapter a description of laminate behavior will be developed using the properties of the lamina and certain assumptions regarding the behavior of the laminate.

3.2 REVIEW OF LAMINA ELASTIC BEHAVIOR

As developed in the previous chapter, an orthotropic layer in a state of plane stress may be characterized by the following constitutive relationship:

$$\begin{bmatrix} \sigma_1 \\ \sigma_2 \\ \tau_{12} \end{bmatrix} = \begin{bmatrix} Q_{11} & Q_{12} & 0 \\ Q_{12} & Q_{22} & 0 \\ 0 & 0 & Q_{66} \end{bmatrix} \begin{bmatrix} \varepsilon_1 \\ \varepsilon_2 \\ \gamma_{12} \end{bmatrix} \tag{3-1}$$

where the stresses σ_1, σ_2, τ_{12}, and strains ε_1, ε_2, γ_{12} are referred to the 1-2-3 coordinate system—the principal or specially orthotropic coordinate system. In this form the stiffness terms have previously been given in terms of the engineering plastic properties of the layer, equation (2-28).

Using the transformation equations given previously, equations (2-35), the stress-strain relationships and elastic properties of the layer can be represented in any other axis system x-y-z (z coincident with the three axis):

$$\begin{bmatrix} \sigma_x \\ \sigma_y \\ \tau_{xy} \end{bmatrix} = \begin{bmatrix} \bar{Q}_{11} & \bar{Q}_{12} & \bar{Q}_{16} \\ \bar{Q}_{12} & \bar{Q}_{22} & \bar{Q}_{26} \\ \bar{Q}_{16} & \bar{Q}_{26} & \bar{Q}_{66} \end{bmatrix} \begin{bmatrix} \varepsilon_x \\ \varepsilon_y \\ \gamma_{xy} \end{bmatrix} \tag{3-2}$$

where now $[\bar{Q}]$ is obtained from $[Q]$ with the transformation equations (2-35) for any given orientation θ of x-y with respect to 1-2 (see Figure 2-7). In this chapter the constitutive equations (3-2) for the kth layer will be denoted

$$[\sigma]_k = [\bar{Q}]_k [\varepsilon]_k \quad (3\text{-}3)$$

in the x-y coordinate system.

3.3 STRAIN DISPLACEMENT RELATIONSHIPS

The equations which relate the strain at any point in a laminate undergoing some deformation will be developed in this section in terms of the displacements at the geometrical midplane of the laminate (u_0, v_0) and the displacement in the z direction (w).

Consider a section of a laminate in the x-z plane, Figure 3-1, which is deformed due to some loading. We assume the point A at the geometrical midplane undergoes some displacement u_0 in the x direction. Furthermore, we assume that the normal (perpendicular) to the geometrical midplane, the line BAD, remains straight and normal to the deformed midplane.[1] With this assumption, the displacement of any point on the normal BAD, say of point C, in the x direction, is given by the linear relationship

$$u_c = u_0 - z_c \alpha \quad (3\text{-}4)$$

where, as shown in Figure 3-1, z_c is the z coordinate of the point C measured from the geometrical mid-plane, and α is the slope of the normal BAD with respect to the original vertical line. However, by similar triangles α is also shown to be the slope of the midplane with respect to the z coordinate axis, or

$$\alpha = \frac{\partial w}{\partial x} \quad (3\text{-}5)$$

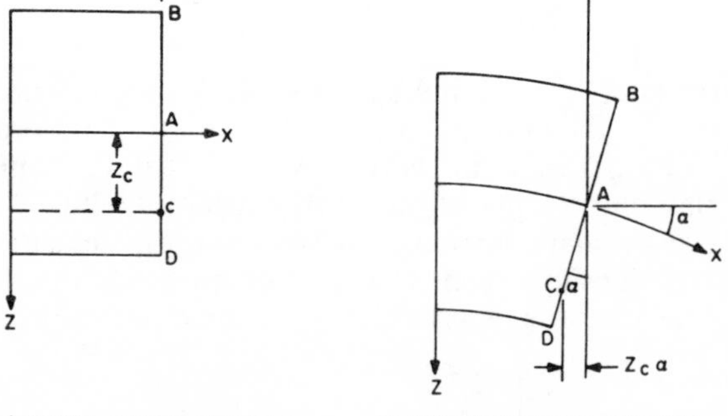

a) INITIAL CROSS-SECTION b) DEFORMED CROSS-SECTION

Figure 3-1. Bending geometry in the x-z plane.

[1]This is equivalent to neglecting shearing deformations γ_{xz} (and γ_{yz}), since these deformations would cause a bending of the normals, and is also equivalent to assuming that the layers that make up the cross-section BAD do not slip over one another.

since the slope is the rate of change of the deflection in the z direction with respect to change in the x direction. Combining equations (3-4) and (3-5), an expression is obtained for the displacement u in the x direction for an arbitrary point at a distance z from the midplane:

$$u = u_0 - z \frac{\partial w}{\partial x} \tag{3-6}$$

In a completely analogous manner, the displacement, v, in the y direction of an arbitrary point at a distance z from the geometrical midplane is

$$v = v_0 - z \frac{\partial w}{\partial y} \tag{3-7}$$

where now v_0 is the midplane displacement in the y direction. Equations (3-6) and (3-7) give the inplane displacement (x, y directions) of any point in the laminate in terms of the inplane displacements at the geometrical midplane (u_0, v_0), and the deflection or displacement w in the z direction.

At this point it is necessary to consider the displacement w further. If the line *BAD* in Figure 3-1 remains straight, then the only displacement in the z direction is the displacement w_0 of the midplane plus the stretching of the normal. It is assumed that the stretching (or shortening) of this line is insignificant, compared to the deflection w, and thus the normal deflection of any point in the laminate is taken equal to the deflection w_0 of the corresponding point at the midplane. Thus the normal strains, ε_z, are neglected when considering normal deflection, and w represents the normal deflection without regard to the z coordinate.

The strains ε_x, ε_y, γ_{xy} at any point in a laminate undergoing some deformation can now be determined in terms of the displacements u_0, v_0, w. The strains γ_{xz}, γ_{yz}, and ε_z need not be considered further since they have been assumed to be negligible in the development above. The normal strain in the x direction will be considered first. Recalling that normal strain is defined to be the change in length divided by the original length, and also recalling that the change in length of a line segment is equal to the relative displacement of the two ends, the normal strain is

$$\varepsilon_x = limit_{\Delta x \to 0} \frac{\Delta u}{\Delta x} = \frac{\partial u}{\partial x} \tag{3-8}$$

Now using equations (3-6) and (3-8), an equation relating normal strain to the laminate displacements (a strain-displacement equation) is obtained:

$$\varepsilon_x = \frac{\partial u}{\partial x} = \frac{\partial}{\partial x}\left(u_0 - z \frac{\partial w}{\partial x}\right) = \frac{\partial u_0}{\partial x} - z \frac{\partial^2 w}{\partial x^2} \tag{3-9}$$

In an analogous manner, the normal strain in the y direction, ε_y, for an arbitrary point in the laminate can be written as follows:

$$\varepsilon_y = \frac{\partial v}{\partial y} = \frac{\partial}{\partial y}\left(v_0 - z\,\frac{\partial w}{\partial y}\right) = \frac{\partial v_0}{\partial y} - z\,\frac{\partial^2 w}{\partial y^2} \tag{3-10}$$

The shear strain of an arbitrary point is defined to be the total change in angle (Figure 3-2) and thus can be written (for small changes of angle)

$$\gamma_{xy} = limit_{\substack{\Delta x \to 0 \\ \Delta y \to 0}}\left(\frac{\Delta u}{\Delta y} + \frac{\Delta v}{\Delta x}\right) = \frac{\partial u}{\partial y} + \frac{\partial v}{\partial x} \tag{3-11}$$

Using now equations (3-6) and (3-7) with equations (3-11) the shear strain can be related to the laminate displacements:

$$\gamma_{xy} = \frac{\partial}{\partial y}\left(u_0 - z\,\frac{\partial w}{\partial x}\right) + \frac{\partial}{\partial x}\left(v_0 - z\,\frac{\partial w}{\partial y}\right) = \frac{\partial u_0}{\partial y} + \frac{\partial v_0}{\partial x} - 2z\,\frac{\partial^2 w}{\partial x \partial y} \tag{3-12}$$

Equations (3-9), (3-10), and (3-12) are the strain-displacement equations for the general (small) deformation of a laminate. These relationships can be written in terms of the midplane strains ε^0_x, ε^0_y, γ^0_{xy} and the plate curvatures k_x, k_y, k_{xy}, as follows:

$$\begin{aligned} \varepsilon^0_x &= \frac{\partial u_0}{\partial x} & k_x &= -\frac{\partial^2 w}{\partial x^2} \\ \varepsilon^0_y &= \frac{\partial v_0}{\partial y} & k_y &= -\frac{\partial^2 w}{\partial y^2} \\ \gamma^0_{xy} &= \frac{\partial u_0}{\partial y} + \frac{\partial v_0}{\partial x} & k_{xy} &= -2\,\frac{\partial^2 w}{\partial x \partial y} \end{aligned} \tag{3-13}$$

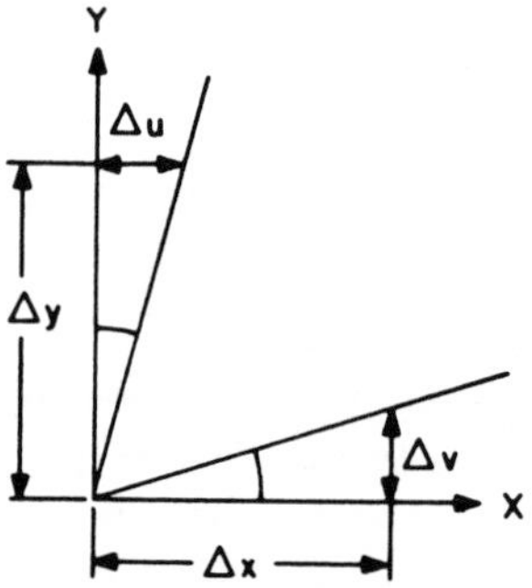

Figure 3-2. Shear strain.

The above expressions for the midplane strains are obtained by simply applying the definitions for strains to the midplane displacements u_0 and v_0. The expressions for curvatures are actually definitions in this case, although they can also be derived by defining curvature to be the negative of the rate of change of slope.

Using the identities (3-13) with equations (3-9), (3-10), and (3-11), the strain-displacement equations can now be written:

$$\begin{bmatrix} \varepsilon_x \\ \varepsilon_y \\ \gamma_{xy} \end{bmatrix} = \begin{bmatrix} \varepsilon_x^0 \\ \varepsilon_y^0 \\ \gamma_{xy}^0 \end{bmatrix} + z \begin{bmatrix} k_x \\ k_y \\ k_{xy} \end{bmatrix} \tag{3-14}$$

or

$$[\varepsilon] = [\varepsilon^0] + z\,[k] \tag{3-15}$$

It should be remembered that the midplane strains are known functions of the midplane displacements (u_0, v_0) and the curvatures are known functions of the deflection (w).

With equations (3-3) and (3-14) or (3-15), the stress state in the *k*th ply can be written in terms of the midplane strains, the plate curvatures, the z coordinate, and the plate elastic (stiffness) properties:

$$\begin{bmatrix} \sigma_x \\ \sigma_y \\ \tau_{xy} \end{bmatrix}_k = \begin{bmatrix} \bar{Q}_{11} & \bar{Q}_{12} & \bar{Q}_{16} \\ \bar{Q}_{12} & \bar{Q}_{22} & \bar{Q}_{26} \\ \bar{Q}_{16} & \bar{Q}_{26} & \bar{Q}_{66} \end{bmatrix}_k \begin{bmatrix} \varepsilon_x^0 \\ \varepsilon_y^0 \\ \gamma_{xy}^0 \end{bmatrix} + z \begin{bmatrix} \bar{Q}_{11} & \bar{Q}_{12} & \bar{Q}_{16} \\ \bar{Q}_{12} & \bar{Q}_{22} & \bar{Q}_{26} \\ \bar{Q}_{16} & \bar{Q}_{26} & \bar{Q}_{66} \end{bmatrix}_k \begin{bmatrix} k_x \\ k_y \\ k_{xy} \end{bmatrix} \tag{3-16}$$

or

$$[\sigma]_k = [\bar{Q}]_k\,[\varepsilon^0] + z\,[\bar{Q}]_k\,[k] \tag{3-17}$$

This expression can be used to compute the stress in a lamina when the laminate midplane strains and curvatures are known (see section 3-9).

3.4 DEFINITION OF STRESS AND MOMENT RESULTANTS

Because the stresses from layer to layer in a laminate vary, and because it will be convenient to deal with a simpler but equivalent system of forces and moments on a laminate cross-section, definitions are introduced below for stress resultants and moment resultants on an elemental parallelopiped of a laminate. These stress and moment resultants consist of three quantities with dimensions of force per unit length (the stress resultants) and three quantities with dimensions of length times force per length (the moment resultants). Together, these six quantities form a system that is statically equivalent to the stress system on the laminate, but which is applied at the geometrical midplane. In terms of these resultants, the total system of loads acting on a cross-section is thus reduced to a system which does not contain the laminate thickness nor z coordinate explicitly (although, of course, the thickness and z coordinate enter into the definitions of these quantities).

An elemental parallelopiped cut from a laminate is shown in Figure 3-3. In the general case such a parallelopiped has stresses σ_x, σ_y, σ_z, τ_{xy}, τ_{xz}, and τ_{yx} acting on the faces. In light of our previous assumptions regarding the transverse shear and the normal stress in the z direction, only σ_x, σ_y, and τ_{xy} will be considered since the laminate is in a plane stress state. In order to find a system of forces and moments acting at the geometric midplane that is equivalent to the effect of these stresses, three stress resultants are defined which are equal to the sum or integral of these stresses in the z direction:

$$
\begin{aligned}
N_x &= \int_{-h/2}^{h/2} \sigma_x dz \\
N_y &= \int_{-h/2}^{h/2} \sigma_y dz \qquad (3\text{-}18) \\
N_{xy} &= \int_{-h/2}^{h/2} \tau_{xy} dz
\end{aligned}
$$

Where the average stress resultants $\sigma_i = N_i/h$.

These stress resultants are positive in the same sense as the corresponding stresses, and since these quantities are stresses times length *(dz)*, they have the dimension of force per length. They are the total load acting per unit length at the midplane.

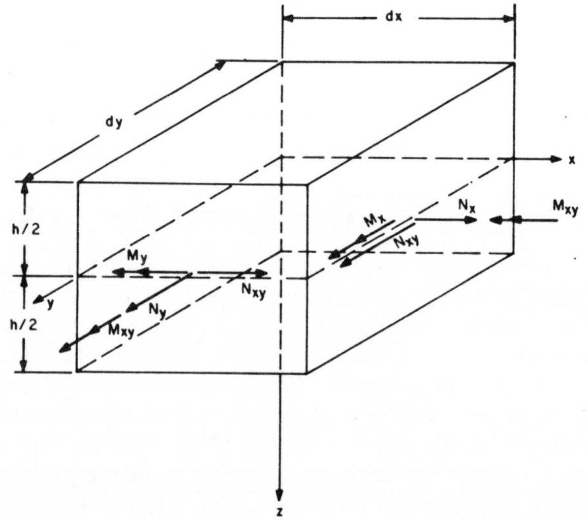

Figure 3-3. *Stress and moment resultants on elemental parallelopiped.*

The resultant of the stresses is not given entirely by the equivalent total loads. In addition, moments must be applied at the midplane which are statically equivalent to the moments created by the stresses with respect to this midplane. These moment resultants are given as the sum of the stresses times the area over which they act multiplied by the moment arm with respect to the midplane, or, in terms of unit lengths along the midplane:

$$\begin{aligned} M_x &= \int_{-h/2}^{h/2} \sigma_x z \, dz \\ M_y &= \int_{-h/2}^{h/2} \sigma_y z \, dz \\ M_{xy} &= \int_{-h/2}^{h/2} \tau_{xy} z \, dz \end{aligned} \tag{3-19}$$

The positive senses of the moments in terms of the moment vector are shown in Figure 3-3 (shown as a double arrow following the right hand rule). The moment resultants give the moment per length at the midplane.

With the definitions of equations (3-18) and (3-19), a system of three stress resultants and three moment resultants has been found which is statically equivalent to the actual stresses distributed across the thickness of the laminate.

3.5 LAMINATE CONSTITUTIVE EQUATIONS [3-1]

Equations (3-18) and (3-19) define a force and moment system acting at the midplane of a laminate in terms of the laminate stresses. Equations (3-16) or (3-17) define the stresses acting on any layer or lamina in terms of the midplane strains (which are functions of the midplane displacements) and the plate curvatures (which are functions of the deflection w). By combining these equations, relationships between the force and moment system, the midplane strains, and the plate curvatures can be obtained. These relationships are the laminate constitutive equations.

The stress resultants, written in the form of a vector, are given in terms of the stress vector by using the definition equation (3-18):

$$\begin{bmatrix} N_x \\ N_y \\ N_{xy} \end{bmatrix} = \int_{-h/2}^{h/2} \begin{bmatrix} \sigma_x \\ \sigma_y \\ \tau_{xy} \end{bmatrix} dz \tag{3-20}$$

By applying the definitions in this form to equation (3-16), separating the continuous integral into an integral over each of the n layers, using the notation of Figure 3-4, the stress resultants are expressed as the sum of n simple integrals:

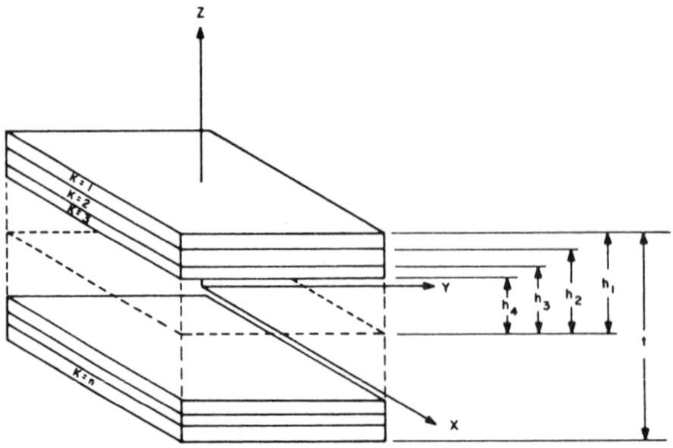

Figure 3-4. Notation for lamina coordinate within a laminate.

$$\begin{bmatrix} N_x \\ N_y \\ N_{xy} \end{bmatrix} = \sum_{k=1}^{n} \int_{h_{k-1}}^{h_k} \begin{bmatrix} \sigma_x \\ \sigma_y \\ \tau_{xy} \end{bmatrix}_k dz \tag{3-21}$$

$$= \sum_{k=1}^{n} \left\{ \int_{h_{k-1}}^{h_k} \begin{bmatrix} \bar{Q}_{11} & \bar{Q}_{12} & \bar{Q}_{16} \\ \bar{Q}_{12} & \bar{Q}_{22} & \bar{Q}_{26} \\ \bar{Q}_{16} & \bar{Q}_{26} & \bar{Q}_{66} \end{bmatrix}_k \begin{bmatrix} \varepsilon_x^0 \\ \varepsilon_y^0 \\ \gamma_{xy}^0 \end{bmatrix} dz \right.$$

$$\left. + \int_{h_{k-1}}^{h_k} \begin{bmatrix} \bar{Q}_{11} & \bar{Q}_{12} & \bar{Q}_{16} \\ \bar{Q}_{12} & \bar{Q}_{22} & \bar{Q}_{26} \\ \bar{Q}_{16} & \bar{Q}_{26} & \bar{Q}_{66} \end{bmatrix}_k \begin{bmatrix} k_x \\ k_y \\ k_{xy} \end{bmatrix} z\, dz \right\}$$

Each of these integrals can be easily evaluated since $[\varepsilon^0]$ and $[k]$ are not functions of z and within any layer (within h_{k-1} to h_k) $[\bar{Q}]$ is not a function of z.
Therefore

$$\begin{bmatrix} N_x \\ N_y \\ N_{xy} \end{bmatrix} = \sum_{k=1}^{n} \left\{ \begin{bmatrix} \bar{Q}_{11} & \bar{Q}_{12} & \bar{Q}_{16} \\ \bar{Q}_{12} & \bar{Q}_{22} & \bar{Q}_{26} \\ \bar{Q}_{16} & \bar{Q}_{26} & \bar{Q}_{66} \end{bmatrix}_k \begin{bmatrix} \varepsilon_x^0 \\ \varepsilon_y^0 \\ \gamma_{xy}^0 \end{bmatrix} \int_{h_{k-1}}^{h_k} dz \right.$$

$$\left. + \begin{bmatrix} \bar{Q}_{11} & \bar{Q}_{12} & \bar{Q}_{16} \\ \bar{Q}_{12} & \bar{Q}_{22} & \bar{Q}_{26} \\ \bar{Q}_{16} & \bar{Q}_{26} & \bar{Q}_{66} \end{bmatrix}_k \begin{bmatrix} k_x \\ k_y \\ k_{xy} \end{bmatrix} \int_{h_{k-1}}^{h_k} z\, dz \right\} \tag{3-22}$$

Furthermore, $[\varepsilon^0]$ and $[k]$ are not functions of k and thus equation (3-22) can be reduced to the following relatively simple form:

$$\begin{bmatrix} N_x \\ N_y \\ N_{xy} \end{bmatrix} = \begin{bmatrix} A_{11} & A_{12} & A_{16} \\ A_{12} & A_{22} & A_{26} \\ A_{16} & A_{26} & A_{66} \end{bmatrix} \begin{bmatrix} \varepsilon_x^0 \\ \varepsilon_y^0 \\ \gamma_{xy}^0 \end{bmatrix} + \begin{bmatrix} B_{11} & B_{12} & B_{16} \\ B_{12} & B_{22} & B_{26} \\ B_{16} & B_{26} & B_{66} \end{bmatrix} \begin{bmatrix} k_x \\ k_y \\ k_{xy} \end{bmatrix} \tag{3-23}$$

or

$$[N] = [A]\,[\varepsilon^0] + [B]\,[k] \tag{3-24}$$

where

$$A_{ij} = \sum_{k=1}^{n} (\bar{Q}_{ij})_k \,(h_k - h_{k-1}) \tag{3-25}$$

$$B_{ij} = \frac{1}{2} \sum_{k=1}^{n} (\bar{Q}_{ij})_k \,(h_k^2 - h_{k-1}^2) \tag{3-26}$$

Equation (3-23) indicates that for a general laminated plate the midplane stress resultants are given in terms of the midplane strains *and* the plate curvatures for *small* normal curvatures. That is, both stretching (or compressing) and bending of the laminate induces midplane stress resultants. Furthermore, in the general case, the normal stress resultants N_x and N_y are developed in part by shearing of the midplane and by twisting of the plate, and the shear stress resultant N_{xy} is developed in part by normal strains of the midplane and normal bending of the plate.

Figure 3-5 shows a two layer strip fabricated of orthotropic layers of nylon reinforced rubber under a tensile load. This tensile specimen is free to rotate and thus only the stress resultant N_x is non-zero, that is, $N_y = N_{xy} = 0$, and the moment resultants M_x, M_y, M_{xy} are also zero. The stacking sequence for this specimen is +30/−30 with respect to the natural axes of the orthotropic rubber layers. The equation for N_x is, in this case:

$$N_x = A_{11}\,\varepsilon_x^0 + A_{12}\,\varepsilon_y^0 + B_{16}k_{xy}$$

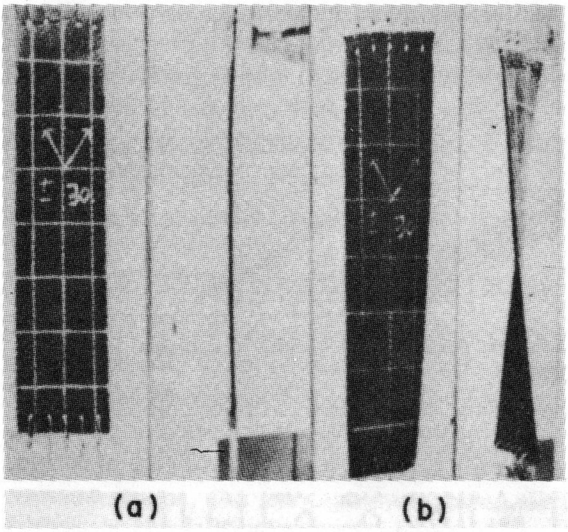

Figure 3-5. *Unsymmetrical laminate under tensile load.*

since A_{16}, B_{11}, B_{12} are zero for this case (see section 3.6). This equation indicates that N_x is given in terms of midplane strains and the twisting of the strip and indeed, as shown in Figure 3-5, the strip twists under a tensile load. The physical source of the B_{16} response is the moment couple developed by the η_{16} response, equation (2-39), acting across a distance as illustrated in Figure 3-6.

The moment resultants can be defined in terms of the stress vector, using equation (3-19):

$$\begin{bmatrix} M_x \\ M_y \\ M_{xy} \end{bmatrix} = \int_{h_{k-1}}^{h_k} \begin{bmatrix} \sigma_x \\ \sigma_y \\ \tau_{xy} \end{bmatrix} z \, dz \tag{3-27}$$

This definition is applied to the stress results of equation (3-16), and again the integration is separated into n sub-intervals:

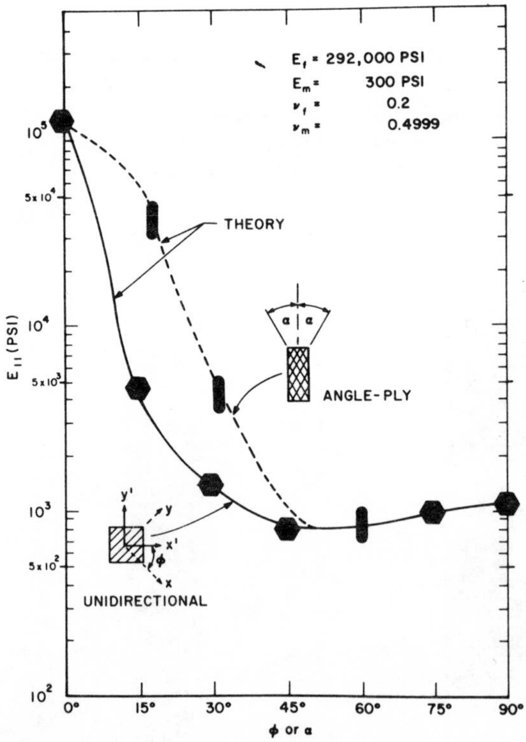

Figure 3-6. Comparison of theory and experiment for unidirectional lamina and angle ply laminates of nylon elastomeric ply material.

$$\begin{bmatrix} M_x \\ M_y \\ M_{xy} \end{bmatrix} = \sum_{k=1}^{n} \int_{h_{k-1}}^{h_k} \begin{bmatrix} \sigma_x \\ \sigma_y \\ \tau_{xy} \end{bmatrix}_k z\, dz \tag{3-28}$$

$$= \sum_{k=1}^{n} \left\{ \int_{h_{k-1}}^{h_k} \begin{bmatrix} \bar{Q}_{11} & \bar{Q}_{12} & \bar{Q}_{16} \\ \bar{Q}_{12} & \bar{Q}_{22} & \bar{Q}_{26} \\ \bar{Q}_{16} & \bar{Q}_{26} & \bar{Q}_{66} \end{bmatrix}_k \begin{bmatrix} \varepsilon_x^0 \\ \varepsilon_y^0 \\ \gamma_{xy}^0 \end{bmatrix} z\, dz \right.$$

$$\left. + \int_{h_{k-1}}^{h_k} \begin{bmatrix} \bar{Q}_{11} & \bar{Q}_{12} & \bar{Q}_{16} \\ \bar{Q}_{12} & \bar{Q}_{22} & \bar{Q}_{26} \\ \bar{Q}_{16} & \bar{Q}_{26} & \bar{Q}_{66} \end{bmatrix}_k \begin{bmatrix} k_x \\ k_y \\ k_{xy} \end{bmatrix} z^2 dz \right\}$$

Following the same procedure as for the stress resultants, the $[\bar{Q}]_k$ matrix can be removed from the integral and $[\varepsilon^0]$ and $[k]$ can be removed from the summation:

$$\begin{bmatrix} M_x \\ M_y \\ M_{xy} \end{bmatrix} = \left\{ \sum_{k=1}^{n} \int_{h_{k-1}}^{h_k} z\, dz \begin{bmatrix} \bar{Q}_{11} & \bar{Q}_{12} & \bar{Q}_{16} \\ \bar{Q}_{12} & \bar{Q}_{22} & \bar{Q}_{26} \\ \bar{Q}_{16} & \bar{Q}_{26} & \bar{Q}_{66} \end{bmatrix}_k \begin{bmatrix} \varepsilon_x^0 \\ \varepsilon_y^0 \\ \gamma_{xy}^0 \end{bmatrix} \right\}$$

$$+ \left\{ \sum_{k=1}^{n} \int_{h_{k-1}}^{h_k} z^2\, dz \begin{bmatrix} \bar{Q}_{11} & \bar{Q}_{12} & \bar{Q}_{16} \\ \bar{Q}_{12} & \bar{Q}_{22} & \bar{Q}_{26} \\ \bar{Q}_{16} & \bar{Q}_{26} & \bar{Q}_{66} \end{bmatrix}_k \begin{bmatrix} k_x \\ k_y \\ k_{xy} \end{bmatrix} \right\} \tag{3-29}$$

Carrying out the indicated integrations, the constitutive relationship for the moment resultant is obtained:

$$\begin{bmatrix} M_x \\ M_y \\ M_{xy} \end{bmatrix} = \begin{bmatrix} B_{11} & B_{12} & B_{16} \\ B_{12} & B_{22} & B_{26} \\ B_{16} & B_{26} & B_{66} \end{bmatrix} \begin{bmatrix} \varepsilon_x^0 \\ \varepsilon_y^0 \\ \gamma_{xy}^0 \end{bmatrix} + \begin{bmatrix} D_{11} & D_{12} & D_{16} \\ D_{12} & D_{22} & D_{26} \\ D_{16} & D_{26} & D_{66} \end{bmatrix} \begin{bmatrix} k_x \\ k_y \\ k_{xy} \end{bmatrix} \tag{3-30}$$

or

$$[M] = [B]\,[\varepsilon^0] + [D]\,[k] \tag{3-31}$$

where

$$B_{ij} = \frac{1}{2} \sum_{k=1}^{n} (Q_{ij})_k \,(h_k^2 - h_{k-1}^2) \tag{3-26}$$

$$D_{ij} = \frac{1}{3} \sum_{k=1}^{n} (Q_{ij})_k \,(h_k^3 - h_{k-1}^3) \tag{3-32}$$

Equation (3-30) indicates that for a general laminated plate the bending moments arise or are given in terms of the midplane strains and the plate curvatures. That is, stretching (or compressing) the midplane, as well as enforcing curvatures, results in bending moments. Furthermore, the normal bending moments M_x and M_y are developed in part due to midplane shearing and plate twisting, and the twisting moment M_{xy} is developed partly due to midplane normal strains and normal plate curvatures.

Combining equations (3-24) and (3-31), the total plate constitutive equation can be written as follows:

$$\left[\frac{N}{M}\right] = \left[\begin{array}{c|c} A & B \\ \hline B & D \end{array}\right] \left[\frac{\varepsilon^0}{k}\right] \tag{3-33}$$

As noted above, the constitutive equations are coupled between bending and stretching. In addition, coupling between normal stress, shearing, and twisting deformations is exhibited. To clarify this result, consider the expression for the normal stress resultant N_x:

$$N_x = A_{11}\varepsilon_x^0 + A_{12}\varepsilon_y^0 + A_{16}\gamma_{xy}^0 + B_{11}k_x + B_{12}k_y + B_{16}k_{xy} \tag{3-34}$$

This stress resultant is a function of the stretching of the middle surface (ε_x^0, ε_y^0). Additionally, a part of N_x is contributed by the shearing of the middle surface (γ_{xy}^0) and by the bending and twisting of the laminate (k_x, k_y, k_{xy}). If $B_{11} = B_{12} = B_{16} = 0$, then

$$N_x = A_{11}\varepsilon_x^0 + A_{12}\varepsilon_y^0 + A_{16}\gamma_{xy}^0 \tag{3-35}$$

and the normal stress resultant in the x direction becomes only a function of the mid-plane stretching and shearing. If in addition $A_{16} = 0$,

$$N_x = A_{11}\varepsilon_x^0 + A_{12}\varepsilon_y^0 \tag{3-36}$$

and the normal stress resultant is a function of only the stretching of the midplane. Equation (3-36) is of the same form as the normal stress-midplane strain relationship for homogeneous specially orthotropic plates. An equation of this form also gives the relations for the more familiar homogeneous isotropic plates.

In an analogous manner, the normal bending moment M_x can be reduced in complexity from the most general case

$$M_x = B_{11}\varepsilon_x^0 + B_{12}\varepsilon_y^0 + B_{16}\gamma_{xy}^0 + D_{11}k_x + D_{12}k_y + D_{16}k_{xy} \tag{3-37}$$

in which the midplane strains contribute to the normal bending moment, to

$$M_x = D_{11}k_x + D_{12}k_y + D_{16}k_{xy} \tag{3-38}$$

when $B_{11} = B_{12} = B_{16} = 0$. In this equation, the normal bending moment M_x

is a function of the normal curvatures and also the twisting of the laminate. If in addition $D_{16} = 0$, then

$$M_x = D_{11}k_x + D_{12}k_y \tag{3-39}$$

and the normal moment is a function of only the normal curvatures of the laminate. Equation (3-39) is also of the same form as the normal moment-curvature relationship for a homogeneous specially orthotropic plate. An equation of this form also gives the relations for the more familiar homogeneous isotropic plates.

The above specific discussion concerning N_x and M_x can be extended to the total set of constitutive equations (3-33). The general case, involving bending-stretching coupling, is written in full for clarity:

$$\begin{bmatrix} N_x \\ N_y \\ N_{xy} \\ \hline M_x \\ M_y \\ M_{xy} \end{bmatrix} = \left[\begin{array}{ccc|ccc} A_{11} & A_{12} & A_{16} & B_{11} & B_{12} & B_{16} \\ A_{12} & A_{22} & A_{26} & B_{12} & B_{22} & B_{26} \\ A_{16} & A_{26} & A_{66} & B_{16} & B_{26} & B_{66} \\ \hline B_{11} & B_{12} & B_{16} & D_{11} & D_{12} & D_{16} \\ B_{12} & B_{22} & B_{26} & D_{12} & D_{22} & D_{26} \\ B_{16} & B_{26} & B_{66} & D_{16} & D_{26} & D_{66} \end{array}\right] \begin{bmatrix} \varepsilon_x^0 \\ \varepsilon_y^0 \\ \gamma_{xy}^0 \\ \hline k_x \\ k_y \\ k_{xy} \end{bmatrix} \tag{3-40}$$

Example 3-1

To illustrate the calculation of the constitutive matrix for a general laminate, a two-ply 0/+45° laminate is considered. The bottom lamina of this laminate is a 0° layer with the following properties

$$\begin{bmatrix} \sigma_1 \\ \sigma_2 \\ \tau_{12} \end{bmatrix}_{0°} = \begin{bmatrix} \sigma_x \\ \sigma_y \\ \tau_{xy} \end{bmatrix}_{0°} = 10^6 \begin{bmatrix} 30. & 1. & 0. \\ 1. & 3. & 0. \\ 0. & 0. & 1. \end{bmatrix} \begin{bmatrix} \varepsilon_1 \\ \varepsilon_2 \\ \gamma_{12} \end{bmatrix} = 10^6 \begin{bmatrix} 30. & 1. & 0. \\ 1. & 3. & 0. \\ 0. & 0. & 1. \end{bmatrix} \begin{bmatrix} \varepsilon_x \\ \varepsilon_y \\ \gamma_{xy} \end{bmatrix}$$

Since this is a zero degree ply, the x-y coordinate system coincides with the 1-2 coordinate system. This layer is .20″ thick.

The second lamina is a +45 layer with the same properties as the first lamina (in the 1-2 coordinate system), and it is .10″ thick. The Q_{ij} terms are found in the x-y coordinate system by using the transformation equations (2-35):

$$\begin{aligned} \bar{Q}_{11} &= 10^6[30 \cos^4(45) + 2.(1. + 2.) \sin^2(45) \cos^2(45) + 3. \sin^4(45)] \\ &= 9.75 \times 10^6 \\ \bar{Q}_{12} &= 10^6[(30. + 3. - 4.) \sin^2(45) \cos^2(45) + 1.(\sin^4(45) + \cos^4(45))] \\ &= 7.75 \times 10^6 \\ \bar{Q}_{22} &= 10^6[30. \sin^4(45) + 2.(1. + 2.) \sin^2(45) \cos^2(45) + 3. \cos^4(45)] \\ &= 9.75 \times 10^6 \\ \bar{Q}_{66} &= 10^6[(30. + 3. - 2. - 2.) \sin^2(45) \cos^2(45) + 1.(\sin^4(45) + \cos^4(45))] \\ &= 7.75 \times 10^6 \end{aligned}$$

$$\bar{Q}_{16} = 10^6[(30. - 1. - 2.)\sin(45)\cos^3(45) + (1. - 3. + 2.)\sin^3(45)\cos(45)] = 6.75 \times 10^6$$

$$\bar{Q}_{26} = 10^6[(30. - 1. - 2.)\sin^3(45)\cos(45) + (1. - 3. + 2.)\sin(45)\cos^3(45)] = 6.75 \times 10^6$$

Therefore

$$\begin{bmatrix} \sigma_x \\ \sigma_y \\ \tau_{xy} \end{bmatrix}_{+45°} = 10^6 \begin{bmatrix} 9.75 & 7.75 & 6.75 \\ 7.75 & 9.75 & 6.75 \\ 6.75 & 6.75 & 7.75 \end{bmatrix} \begin{bmatrix} \varepsilon_x \\ \varepsilon_y \\ \gamma_{xy} \end{bmatrix}$$

Now the basic terms in the definitions (3-25), (3-26), and (3-32) are known ($h_0 = -.15$, $h_1 = .05$, $h_2 = .15$ from the given lamina dimensions). Applying these definitions, the laminate constitutive equation is obtained as follows:

$$A_{ij} = \sum_{k=1}^{n} (\bar{Q}_{ij})_k (h_k - h_{k-1}) = (\bar{Q}_{ij})_1 [.05-(-.15)] + (\bar{Q}_{ij})_2 [.15-.05]$$
$$= .2(\bar{Q}_{ij})_1 + .1(\bar{Q}_{ij})_2$$

Thus

$$A_{11} = 10^6[30. \cdot .2 + 9.75 \cdot .1] = 6.975 \times 10^6$$
$$A_{12} = 10^6[1. \times .2 + 7.75 \times .1] = .975 \times 10^6$$
$$\vdots \qquad \qquad \vdots$$
$$A_{66} = 10^6[1. \times .2 + 7.75 \times .1] = .975 \times 10^6$$

$$[A] = 10^6 \begin{bmatrix} 6.975 & .975 & .675 \\ .975 & 1.575 & .675 \\ .675 & .675 & .975 \end{bmatrix}$$

$$B_{ij} = \frac{1}{2}\sum_{k=1}^{n} (\bar{Q}_{ij})_k (h_k^2 - h_{k-1}^2) = \frac{1}{2}(\bar{Q}_{ij})_1 [.05^2-(-.15)^2]$$
$$+ \frac{1}{2}(\bar{Q}_{ij})_2 [.15^2-.05^2] = .01\,[(\bar{Q}_{ij})_2-(\bar{Q}_{ij})_1]$$

Thus

$$B_{11} = 10^6 \times .01\,[9.75 - 30.] = -20.25 \times 10^4$$
$$B_{12} = 10^6 \times .01\,[7.75 - 1.] = 6.75 \times 10^4$$
$$\vdots \qquad \qquad \vdots$$
$$B_{66} = 10^6 \times .01\,[7.75 - 1.] = 6.75 \times 10^4$$

$$[B] = 10^4 \begin{bmatrix} -20.25 & 6.75 & 6.75 \\ 6.75 & 6.75 & 6.75 \\ 6.75 & 6.75 & 6.75 \end{bmatrix}$$

$$D_{ij} = \frac{1}{3}\sum_{k=1}^{n} (\bar{Q}_{ij})_k \, (h_k^3 - h_{k-1}^3) = \frac{1}{3} (\bar{Q}_{ij})_1 \, [.05^3 - (-.15)^3]$$

$$+ \frac{1}{3} (\bar{Q}_{ij})_2 \, [.15^3 - .05^3] = (\bar{Q}_{ij})_1 (.1167 \times 10^{-2})$$
$$+ (\bar{Q}_{ij})_2 \, (.1083 \times 10^{-2})$$

Thus

$$D_{11} = (30. \times 10^6)(.1167 \times 10^{-2}) + (9.75 \times 10^6)(.1083 \times 10^{-2})$$
$$= 4.56 \times 10^4$$
$$D_{12} = (1. \times 10^6)(.1167 \times 10^{-2}) + (7.75 \times 10^6)(.1083 \times 10^{-2}) = .83 \times 10^4$$

$$\vdots \qquad\qquad \vdots$$

$$D_{66} = (1. \times 10^6)(.1167 \times 10^{-2}) + (7.75 \times 10^6)(.1083 \times 10^{-2}) = .83 \times 10^4$$

$$[D] = 10^4 \begin{bmatrix} 4.56 & .83 & .54 \\ .83 & 1.41 & .54 \\ .54 & .54 & .83 \end{bmatrix}$$

Combining the above results, the total set of constitutive equations for this particular two-ply laminate can be written:

$$\begin{bmatrix} N_x \\ N_y \\ N_{xy} \\ \hline M_x \\ M_y \\ M_{xy} \end{bmatrix} = 10^6 \left[\begin{array}{ccc|ccc} 697.5 & 97.5 & 67.5 & -20.25 & 6.75 & 6.75 \\ 97.5 & 157.5 & 67.5 & 6.75 & 6.75 & 6.75 \\ 67.5 & 67.5 & 97.5 & 6.75 & 6.75 & 6.75 \\ \hline -20.25 & 6.75 & 6.75 & 4.56 & .83 & .54 \\ 6.75 & 6.75 & 6.75 & .83 & 1.41 & .54 \\ 6.75 & 6.75 & 6.75 & .54 & .54 & .83 \end{array}\right] \begin{bmatrix} \varepsilon_x^0 \\ \varepsilon_y^0 \\ \gamma_{xy}^0 \\ \hline k_x \\ k_y \\ k_{xy} \end{bmatrix}$$

The bending-membrane coupling introduced by the *[B]* matrix is not due to the orthotropy or anisotropy of the layers, but rather is due to the stacking sequence or the heterogenity of the laminate. In fact, this coupling exists for some isotropic laminates, as illustrated in the following example.

Example 3-2

We consider a two-ply laminate. The bottom layer is aluminum and thus has the following constitutive properties ($E = 10^7$, $\nu = .33$):

$$\begin{bmatrix} \sigma_1 \\ \sigma_2 \\ \tau_{12} \end{bmatrix} = 10^6 \begin{bmatrix} 11.25 & 3.73 & 0. \\ 3.73 & 11.25 & 0. \\ 0. & 0. & 3.75 \end{bmatrix} \begin{bmatrix} \varepsilon_1 \\ \varepsilon_2 \\ \gamma_{12} \end{bmatrix}$$

The second layer is steel and has the constitutive properties given below ($E = 29 \times 10^6$, $\nu = .28$):

$$\begin{bmatrix} \sigma_1 \\ \sigma_2 \\ \tau_{12} \end{bmatrix} = 10^6 \begin{bmatrix} 31.6 & 8.9 & 0. \\ 8.9 & 31.6 & 0. \\ 0. & 0. & 11.3 \end{bmatrix} \begin{bmatrix} \varepsilon_1 \\ \varepsilon_2 \\ \gamma_{12} \end{bmatrix}$$

Both layers are 1.0″ thick. Since the layers are isotropic, any x-y coordinate system will have the same constitutive equations as those given above. Applying the definitions (3-25), (3-26), and (3-32) with $h_0 = -1.$, $h_1 = 0.$, and $h_2 = 1.$, the following results are obtained:

$$\begin{aligned} A_{ij} &= (\bar{Q}_{ij})_1\,(0. + 1.) + (\bar{Q}_{ij})_2\,(1. - 0.) = (\bar{Q}_{ij})_1 + (\bar{Q}_{ij})_2 \\ 2B_{ij} &= (\bar{Q}_{ij})_1\,(0^2_. - 1^2_.) + (\bar{Q}_{ij})_2\,(1^2_. - 0^2_.) = (\bar{Q}_{ij})_2 - (\bar{Q}_{ij})_1 \\ 3D_{ij} &= (\bar{Q}_{ij})_1\,(0^3_. - (-1.)^3) + (\bar{Q}_{ij})_2\,(1^3_. - 0^3_.) = (\bar{Q}_{ij})_1 + (\bar{Q}_{ij})_2 \end{aligned}$$

or

$$\begin{bmatrix} N_x \\ N_y \\ N_{xy} \\ \hline M_x \\ M_y \\ M_{xy} \end{bmatrix} = 10^6 \left[\begin{array}{ccc|ccc} 42.85 & 12.63 & 0 & 10.18 & 2.58 & 0. \\ 12.63 & 42.85 & 0. & 2.58 & 10.18 & 0. \\ 0. & 0 & 15.05 & 0. & 0. & 3.78 \\ \hline 10.18 & 2.58 & 0. & 14.28 & 4.21 & 0. \\ 2.58 & 10.18 & 0. & 4.21 & 14.28 & 0. \\ 0. & 0. & 3.78 & 0. & 0. & 5.02 \end{array}\right] \begin{bmatrix} \varepsilon_x^0 \\ \varepsilon_y^0 \\ \gamma_{xy}^0 \\ \hline k_x \\ k_y \\ k_{xy} \end{bmatrix}$$

Note that even though both layers of this laminate are isotropic, the $[B]$ matrix is not identically zero. Thus the bending and stretching behavior of this laminate are coupled. This is caused by the heterogeneous nature of the laminate.

If now each $B_{ij} = 0$, then a major simplification and uncoupling occurs in the equations, and the inplane or stretching problem is uncoupled from the bending problem:

$$\begin{bmatrix} N_x \\ N_y \\ N_{xy} \end{bmatrix} = \begin{bmatrix} A_{11} & A_{12} & A_{16} \\ A_{12} & A_{22} & A_{26} \\ A_{16} & A_{26} & A_{66} \end{bmatrix} \begin{bmatrix} \varepsilon_x^0 \\ \varepsilon_y^0 \\ \gamma_{xy}^0 \end{bmatrix} \tag{3-41}$$

$$\begin{bmatrix} M_x \\ M_y \\ M_{xy} \end{bmatrix} = \begin{bmatrix} D_{11} & D_{12} & D_{16} \\ D_{12} & D_{22} & D_{26} \\ D_{16} & D_{26} & D_{66} \end{bmatrix} \begin{bmatrix} k_x \\ k_y \\ k_{xy} \end{bmatrix} \tag{3-42}$$

These equations are the governing constitutive equations for the (uncoupled) inplane problem and the bending problem. Both equations (3-41) and (3-42) describe constitutive equations that are equivalent to some homogeneous anisotropic medium.

A further simplification is obtained if the A_{16} and A_{26} terms in equation (3-41) are zero. In this case equation (3-41) reduces to

$$\begin{bmatrix} N_x \\ N_y \\ N_{xy} \end{bmatrix} = \begin{bmatrix} A_{11} & A_{12} & 0 \\ A_{12} & A_{22} & 0 \\ 0 & 0 & A_{66} \end{bmatrix} \begin{bmatrix} \varepsilon_x^0 \\ \varepsilon_y^0 \\ \gamma_{xy}^0 \end{bmatrix} \tag{3-43}$$

which is the constitutive equation governing the inplane problem for a specially orthotropic medium. Furthermore, if

$$\left.\begin{aligned} A_{11} &= A_{22} = \left(\frac{E}{1-\nu^2}\right) h \\ A_{12} &= \nu A_{11} \\ A_{66} &= \left(\frac{E}{2(1+\nu)}\right) h \end{aligned}\right\} \tag{3-44}$$

where E = Young's modulus, ν Poisson's ratio, and h = total laminate thickness, then equation (3-43) is the constitutive equation governing the in-plane problem for a homogeneous isotropic medium.

Similarly, if $D_{16} = D_{26} = 0$, equation (3-42) reduces to the constitutive equation governing the bending of a specially orthotropic medium:

$$\begin{bmatrix} M_x \\ M_y \\ M_{xy} \end{bmatrix} = \begin{bmatrix} D_{11} & D_{12} & 0 \\ D_{12} & D_{22} & 0 \\ 0 & 0 & D_{66} \end{bmatrix} \begin{bmatrix} k_x \\ k_y \\ k_{xy} \end{bmatrix} \tag{3-45}$$

And if

$$\left.\begin{aligned} D_{11} &= D_{22} = \frac{Eh_3}{12(1-\nu^2)} \\ D_{12} &= \nu D_{11} \\ D_{66} &= \frac{Eh^3}{24(1+\nu)} \end{aligned}\right\} \tag{3-46}$$

then equation (3-45) is the governing constitutive equation for the bending of a homogeneous isotropic plate.

The effective engineering properties for balanced and symmetric laminates are determined from the stiffness constants, A_{ij}

Longitudinal Young's Modulus:

$$E_x = (A_{11}A_{22} - A_{12}^2)/A_{22} \tag{3-47}$$

Transverse Young's Modulus:

$$E_y = (A_{11}A_{22} - A_{12}^2)/A_{11} \tag{3-48}$$

Longitudinal Poisson's Ratio

$$\nu_{xy} = A_{12}/A_{22} \tag{3-49}$$

Shear Modulus

$$G_{xy} = A_{66} \tag{3-50}$$

3.6 SOURCE OF SIMPLIFICATIONS OF LAMINATE CONSTITUTIVE EQUATIONS [3-2]

The discussion above has indicated that major simplifications are possible in the governing constitutive equations for laminates when (1) *[B]* is identically zero and (2) the "16, 26" terms are zero. In this section, the conditions necessary for such simplifications are discussed.

The terms in the *[B]* matrix are obtained as a sum of terms involving the $[\bar{Q}]$ matrices and squares of the z coordinates of the top and bottom of each ply. Since the B_{ij} are thus even functions of the h_k, they are zero for laminates which are symmetrical with respect to z. That is, each term B_{ij} is zero if for each lamina above the mid-plane there is an identical ply (in properties and orientation) located at the same distance below the mid-plane. Such mid-plane symmetric laminates are an important class of laminates. They are commonly constructed because the governing constitutive equations are considerably simplified, and thus the laminates can be more easily analyzed. In addition, and most importantly, they are desirable because the bending-stretching coupling present in non-symmetric laminates causes undesirable warping due to inplane loads, which may be induced by thermal contrations. With mid-plane symmetric laminates, thermal contractions introduce mid-plane strains (and inplane loads), but they do not introduce bending.

Example 3-3

To illustrate that the B_{ij} are all identically zero for a laminate with mid-plane symmetry, consider a three-ply +45/0/+45 laminate. Let each lamina have the same properties as the lamina considered in example (3-1), let the top and bottom layers have a thickness of .1″ each, and the middle 0° layer have a thickness of .2″. Thus, $h_0 = -.20,\ h_1 = -.10,\ h_2 = .10,\ h_3 = .20.$

The definitions (3-25), (3-26), and (3-32) can now be applied directly. For the B_{ij} terms, using equation (3-26) we have

$$B_{ij} = \frac{1}{2}\sum_{k=1}^{3}(\bar{Q}_{ij})_k\,(h_k^2 - h_{k-1}^2) = \frac{1}{2}(\bar{Q}_{ij})_1\,[(-.10)^2 - (-.20)^2]$$
$$+\ \frac{1}{2}(\bar{Q}_{ij})_2[.10^2 - (-.10)^2] + \frac{1}{2}(\bar{Q}_{ij})_3[.20^2 - .10^2]$$

$$= \frac{1}{2} (\bar{Q}_{ij})_1 (-.03) + \frac{1}{2} (\bar{Q}_{ij})_2 (0.) + \frac{1}{2} (\bar{Q}_{ij})_3 (.03)$$

$$= \frac{.03}{2} [(\bar{Q}_{ij})_3 - (\bar{Q}_{ij})_1]$$

But since layer 1 and layer 3 have the same basic properties and the same orientation, $(\bar{Q}_{ij})_1 = (\bar{Q}_{ij})_3$ and each $B_{ij} = 0$.

Applying the definitions (3-25) and (3-32), the $[A]$ and $[D]$ matrices are found to have the following form:

$$[A] = 10^6 \begin{bmatrix} 7.95 & 1.75 & 1.35 \\ 1.75 & 2.55 & 1.35 \\ 1.35 & 1.35 & 1.75 \end{bmatrix}$$

$$[D] = 10^4 \begin{bmatrix} 22.5 & 15.6 & 13.5 \\ 15.6 & 19.8 & 13.5 \\ 13.5 & 13.5 & 15.6 \end{bmatrix}$$

This three-ply laminate does not exhibit bending membrane coupling since the $B_{ij} = 0$ (because the laminate has midplane symmetry), but does exhibit both inplane and bending anisotropy. That is, the A_{16}, A_{26}, D_{16}, and D_{26} terms are all non-zero.

We now consider the possibility of fabricating a laminate that behaves as an orthotropic layer with respect to inplane forces and strains (i.e., $A_{16} = A_{26} = 0$). It is apparent from equation (3-25) that the A_{ij} are equal to the sums of the lamina $\bar{Q}_{ij}$ times the lamina thickness. Thus, the only way an A_{ij} term can be zero is for either all the $\bar{Q}_{ij}$ to be zero or to have some $\bar{Q}_{ij}$ positive and some negative. Since the $\bar{Q}_{ij}$ terms of a lamina are obtained from the Q_{ij} terms (the orthotropic stiffnesses), and due to the form of the transformation equations (2-35), $\bar{Q}_{11}$, $\bar{Q}_{12}$, $\bar{Q}_{22}$, and $\bar{Q}_{66}$ are always positive and greater than zero. Consequently, A_{11}, A_{12}, A_{22}, and A_{66} are always positive and greater than zero. On the other hand, $\bar{Q}_{16}$ and $\bar{Q}_{26}$ are zero for orientations of 0° or 90°, and can be either positive or negative since these terms are defined in terms of odd powers of sin (θ) and cos (θ). In particular, $\bar{Q}_{16}$ $(\bar{Q}_{26})$ for a plus θ rotation is equal in absolute value but of the opposite sign from the $\bar{Q}_{16}$ $(\bar{Q}_{26})$ for a negative θ rotation. Thus, if for every lamina of a plus θ orientation we have another lamina of the same orthotropic properties and thickness with a negative θ orientation, then the laminate is specially orthotropic with respect to inplane forces and strains ($A_{16} = A_{26} = 0$).

Example 3-4

Consider a four layer +45°/+45°/+45°/+45° laminate. Each layer is assumed to have a thickness of .1 in. and the same orthotropic properties as in example (3-1). In this case $[\bar{Q}]$ for the +45 layers is given as

$$[\bar{Q}]_{layers\ 1\ and\ 4} = 10^6 \begin{bmatrix} 9.75 & 7.75 & 6.75 \\ 7.75 & 9.75 & 6.75 \\ 6.75 & 6.75 & 7.75 \end{bmatrix}$$

and for the -45 layers $[\bar{Q}]$ is of the form

$$[\bar{Q}]_{layers\ 2\ and\ 3} = 10^6 \begin{bmatrix} 9.75 & 7.75 & -6.75 \\ 7.75 & 9.75 & -6.75 \\ -6.75 & -6.75 & 7.75 \end{bmatrix}$$

Using the definition of equation (3-25) the $[A]$ matrix is found to be specially orthotropic:

$$\begin{aligned} A_{ij} &= (\bar{Q}_{ij})_1\ (.1) + (\bar{Q}_{ij})_2\ (.1) + (\bar{Q}_{ij})_3\ (.1) + (\bar{Q}_{ij})_4\ (.1) \\ &= (\bar{Q}_{ij})_1\ (.2) + (\bar{Q}_{ij})_2\ (.2) \end{aligned}$$

or

$$(A) = 10^6 \begin{bmatrix} 3.9 & 3.1 & 0. \\ 3.2 & 3.9 & 0. \\ 0. & 0. & 3.1 \end{bmatrix}$$

Since this laminate possesses midplane symmetry, each $B_{ij} = 0$ also.

The D_{ij} terms are defined in terms of the $\bar{Q}_{ij}$ and the difference between the third power of the z coordinate at the layer top and the layer bottom. Since the geometrical contribution $(h_k^3 - h_{k-1}^3)$ is always positive, it follows from the discussion above that D_{11}, D_{12}, D_{22}, and D_{66} are always positive. On the other hand, the D_{16} and D_{26} tersm are zero if all of the lamina are oriented at either 0° or 90° (since each such layer has $\bar{Q}_{16} = \bar{Q}_{26} = 0$). Furthermore, the D_{16} and D_{26} terms are also zero if for every layer oriented at θ at a given distance above the midplane there is an identical layer at the same distance below the midplane oriented at $-\theta$ (since $\bar{Q}_{16}(+\theta) = -\bar{Q}_{16}(-\theta)$, $\bar{Q}_{26}(+\theta) = -\bar{Q}_{26}(-\theta)$, and $(h_k^3 - h_{k-1}^3)$ is the same for both layers). However, this laminate will then not possess midplane symmetry, and the $B_{ij} \neq 0$. In fact, except for all 0°, all 90°, and 0°/90° laminates, the D_{16} and D_{26} terms are not zero for any midplane symmetric laminate.

However, although the D_{16} and D_{26} terms are not zero for any midplane symmetric laminate except those cross-ply cases noted, for angle ply laminates ($\pm\theta$) fabricated of a large number of alternating lamina, these terms do become small. This is because the contribution of the $+\theta$ layers to the D_{16} and D_{26} terms is opposite in sign to the contribution of the $-\theta$ layers, and although they are located at different distances from the midplane, they tend to cancel effects.

Example 3-5

To illustrate the effect of alternating angle laminations, we consider the following eight ply laminates, all of .4″ total thickness:

1) 8 laminae at +45 (equivalent to one lamina of .4″ thickness)
2) +45/+45/−45/−45/−45/−45/+45/+45 (equivalent to four alternating lamina each of thickness .1″)
3) +45/−45/+45/−45/−45/+45/−45/+45

Case 1. In this case the D_{ij} are all given by the following relationship:

$$D_{ij} = \frac{1}{3} \bar{Q}_{ij} (.2^3-(-.2)^3) = .00533\, \bar{Q}_{ij}$$

$$[D] = 10^4 \begin{bmatrix} 5.20 & 4.13 & 3.60 \\ 4.13 & 5.20 & 3.60 \\ 3.60 & 3.60 & 4.13 \end{bmatrix} \qquad \frac{D_{16}}{D_{11}} = .795$$

Case 2. In this case, D_{11}, D_{12}, and D_{66} are the same as for case 1 because $\bar{Q}_{11}$, $\bar{Q}_{12}$, $\bar{Q}_{22}$, and $\bar{Q}_{66}$ are constant through the thickness and the same for both cases.

$$D_{16} = D_{26} = 2\,\frac{10^6}{3}\,[6.75\,(.2^3-.1^3) - 6.75\,(.1^3-0.^3)] = 2.70 \times 10^4$$

$$[D] = 10^4 \begin{bmatrix} 5.20 & 4.13 & 2.70 \\ 4.13 & 5.20 & 2.70 \\ 2.70 & 2.70 & 4.13 \end{bmatrix} \qquad \frac{D_{16}}{D_{11}} = .520$$

Case 3. D_{11}, D_{12}, D_{22}, and D_{66} are again the same as in the above cases.

$$D_{16} = D_{26} = 2\,\frac{10^6}{3}\,[6.75\,(.2^3-.15^3) - 6.75\,(.15^3-.10^3)$$
$$+\ 6.75\,(.10^3-.05^3) - 6.75\,(.05^3-0.^3)] = 1.35 \times 10^4$$

$$[D] = 10^4 \begin{bmatrix} 5.20 & 4.13 & 1.35 \\ 4.13 & 5.20 & 1.35 \\ 1.35 & 1.35 & 4.13 \end{bmatrix} \qquad \frac{D_{16}}{D_{11}} = .26$$

For all of these laminates, $[B] = [0]$ because they possess midplane symmetry. Cases 2 and 3 also have $A_{16} = A_{26} = 0$ because for each lamina at +45 there is one lamina at −45.

Additional values of A_{ij}, B_{ij}, and D_{ij} for a generic graphite/epoxy class material is tabulated in the appendix for laminates of [0/45/−45]s, [0/45/−45/45/−45/0]s, and [0/45/45]s construction. The thickness of each ply is 0.005 inches. This illustration reinforces the observation that while the inplane shear and normal terms are uncoupled for the symmetric and balanced laminate, the twisting bending terms (D_{16}, D_{26}) do not vanish. In summary, for a balanced laminate (equal number of lamina of $+\theta$ deg. and $-\theta$ deg. fiber orientations) the shear coupling terms vanish.

$$A_{16} = A_{26} = 0 \qquad \textit{(if laminate is balanced)}$$

Second, if the laminae within the laminate are positioned symmetrically with respect to the laminate mid-plane, then the coupling terms, B_{ij}, vanish

$$B_{ij} = 0 \; (i, j = 1, 2, 6) \qquad \text{(if the laminate is symmetric)}$$

3.7 OTHER FORMS FOR THE GENERAL CONSTITUTIVE EQUATIONS [3-1]

The general constitutive equation for a laminate can be written, as shown above, in the form

$$\begin{bmatrix} N \\ \hline M \end{bmatrix} = \left[\begin{array}{c|c} A & B \\ \hline B & D \end{array}\right] \begin{bmatrix} \varepsilon^{\circ} \\ \hline k \end{bmatrix} \tag{3-33}$$

It is in general possible to rearrange these equations to other useful forms by partially or totally inverting equation (3-33). Such inverted forms are often useful and will be used in subsequent discussions.

To obtain the other desired forms of the general constitutive equations (3-33), the equations for $[N]$ and for $[M]$ are considered separately.

$$[N] = [A][\varepsilon^0] + [B][k] \tag{3-24}$$
$$[M] = [B][\varepsilon^0] + [D][k] \tag{3-31}$$

Solving equation (3-24) for the midplane strains gives

$$[\varepsilon^0] = [A^{-1}][N] - [A^{-1}][B][k] \tag{3-51}$$

Substituting equation (3-51) into (3-31) and rearranging yields

$$[M] = [B][A^{-1}][N] + (-[B][A^{-1}][B] + [D])[k] \tag{3-52}$$

Combining equations (3-51) and (3-52) in matrix form gives a partially inverted form of the laminate constitutive equations which is used in plate and shell formulations (Chapter 4):

$$\begin{bmatrix} \varepsilon^0 \\ \hline M \end{bmatrix} = \left[\begin{array}{c|c} A^* & B^* \\ \hline C^* & D^* \end{array}\right] \begin{bmatrix} N \\ \hline k \end{bmatrix} \tag{3-53}$$

where

$$[A^*] = [A^{-1}]$$
$$[B^*] = -[A^{-1}][B]$$
$$[C^*] = [B][A^{-1}]$$
$$[D^*] = [D] - [B][A^{-1}][B]$$

Now using the above definitions and solving equations (3-52) for $[k]$ gives

$$[k] = [D^{*-1}][M] - [D^{*-1}][C^*][N] \tag{3-54}$$

Substituting equation (3-54) into (3-51) gives

$$[\varepsilon^0] = [B^*][D^{*-1}][M] + ([A^*] - [B^*][D^{*-1}][C^*])[N] \tag{3-55}$$

Combining equations (3-54) and (3-55) in matrix form now yields the completely inverted forms of the constitutive equations (3-33):

$$\begin{bmatrix} \varepsilon^0 \\ \hline k \end{bmatrix} = \left[\begin{array}{c|c} A' & B' \\ \hline C' & D' \end{array}\right] \begin{bmatrix} N \\ \hline M \end{bmatrix} \tag{3-56}$$

where

$$\begin{aligned} [A'] &= [A^*] - [B^*][D^{*-1}][C^*] \\ [B'] &= [B^*][D^{*-1}] \\ [C'] &= -[D^{*-1}][C^*] \\ [D'] &= [D^{*-1}] \end{aligned}$$

Equations (3-33), (3-53), and (3-56) are the forms of the laminate consitutive relations that are most useful. It is important to note that each form can be obtained from the basic elastic properties of the laminae (from the $\bar{Q}_{ij}$ for each lamina) and the stacking sequence (z coordinates) through the definition equations (3-25), (3-26), (3-32), and certain simple matrix calculations.

3.8 INVARIANT FORMS OF THE *[A], [B],* AND *[D]* MATRICES [3-3]

As shown in Chapter 2, the transformation equations for a single lamina can be expressed in terms of certain invariant combinations of the stiffness, properties $\bar{Q}_{ij}$ and the double angle cosine and sine terms. If the same material is used for each lamina in a laminated composite, then the *[A], [B],* and *[D]* matrices can also be expressed in terms of these same invariants.

From Chapter 2, we know that the $[\bar{Q}]$ matrix for the kth layer can be expressed in terms of the invariants U_1, U_4, U_5 plus the material constants U_2, U_3 and the orientation θ_k:

$$\begin{aligned} \bar{Q}_{11} &= U_1 + U_2 \cos(2\theta_k) + U_3 \cos(4\theta_k) \\ \bar{Q}_{22} &= U_1 - U_2 \cos(2\theta_k) \\ \bar{Q}_{12} &= U_4 - U_3 \cos(4\theta_k) \\ \bar{Q}_{66} &= U_5 - U_3 \cos(4\theta_k) \\ \bar{Q}_{16} &= + \frac{1}{2} U_2 \sin(2\theta_k) + U_3 \sin \\ \bar{Q}_{26} &= + \frac{1}{2} U_2 \sin(2\theta_k) - U_3 \sin \end{aligned} \tag{2-36}$$

If each layer of a laminate is made of the same material, then U_1, U_2, U_3,

U_4, and U_5 are constant for all the layers. Applying the definition equations (3-25), (3-26), and (3-32), invariant results for the *[A]*, *[B]*, and *[D]* matrices can thus be obtained in terms of these material constants U_1, U_2, U_3, U_4, U_5, and summations involving only the orientations θ_k and z coordinates h_k:

$$A_{ij} = \sum_{k=1}^{n} (Q_{ij})_k (h_k - h_{k-1})$$

Hence

$$A_{11} = U_1 \sum_{k=1}^{n} (h_k - h_{k-1}) + U_2 \sum_{k=1}^{n} cos(2\theta_k)(h_k - h_{k-1})$$

$$+ U_3 \sum_{k=1}^{n} cos(4\theta_k)(h_k - h_{k-1}) \tag{3-57}$$

or

$$A_{11} = U_1 V_{0A} + U_2 V_{1A} + U_3 V_{3A} \tag{3-58}$$

where

$$\left.\begin{aligned} V_{0A} &= h \\ V_{1A} &= \sum_{k=1}^{n} cos(2\theta_k)(h_k - h_{k-1}) \\ V_{3A} &= \sum_{k=1}^{n} cos(4\theta_k)(h_k - h_{k-1}) \end{aligned}\right\} \tag{3-59}$$

In an analogous manner we can find expressions for the remaining elements of *[A]* in terms of the properties U_i and the geometrical terms V_{iA}:

$$\left.\begin{aligned} A_{22} &= U_1 V_{0A} - U_2 V_{1A} + U_3 V_{3A} \\ A_{12} &= U_4 V_{0A} - U_3 V_{3A} \\ A_{66} &= U_5 V_{0A} - U_3 V_{3A} \\ A_{16} &= + .5 U_2 V_{2A} + U_3 V_{4A} \\ A_{26} &= + .5 U_2 V_{2A} - U_3 V_{4A} \end{aligned}\right\} \tag{3-60}$$

where in addition we have

$$\left.\begin{aligned} V_{2A} &= \sum_{k=1}^{n} sin(2\theta_k)(h_k - h_{k-1}) \\ V_{4A} &= \sum_{k=1}^{n} sin(4\theta_k)(h_k - h_{k-1}) \end{aligned}\right\} \quad (3\text{-}61)$$

By examining the geometrical terms (3-58) and (3-61), we see that all of the terms except V_{0A} involve a summation of layer thicknesses times double angle cosine and sine functions. For large variations in the θ_k, these terms become smaller with respect to V_{0A} since the trigonometric contribution varies in sign. Consequently, the constant terms (the terms involving V_{0A}) are the limiting values of the A_{ij} for laminates possessing a large number of different θ_k. The transformation characteristics for the in-plane stretching modulus, A_{ij}, of multidirectional symmetrical laminates is:

$$\begin{aligned} \bar{A}_{11} &= U_1 + U_2 \cos 2\theta + U_3 \cos 4\theta \\ \bar{A}_{22} &= U_1 - U_2 \cos 2\theta + U_3 \cos 4\theta \\ \bar{A}_{12} &= U_4 - U_5 \cos 4\theta \\ \bar{A}_{66} &= U_5 - U_3 \cos 4\theta \\ \bar{A}_{16} &= \frac{1}{2} U_2 \sin 2\theta + U_3 \sin 4\theta \\ \bar{A}_{26} &= \frac{1}{2} U_2 \sin 2\theta - U_3 \sin 4\theta \end{aligned} \quad (3\text{-}62)$$

Following precisely the same procedure as outlined above, expressions can be obtained for the *[B]* and *[D]* matrices in terms of geometrical constants and the U_i. These expressions are given below without further derivation.

$$\left.\begin{aligned} B_{11} &= U_2 V_{1B} + U_3 V_{3B} & B_{66} &= -U_3 V_{3B} \\ B_{22} &= -U_2 V_{1B} + U_3 V_{3B} & B_{16} &= +.5U_2 V_{2B} + U_3 V_{4B} \\ B_{12} &= -U_3 V_{3B} & B_{26} &= +.5U_2 V_{2B} - U_3 V_{4B} \end{aligned}\right\} \quad (3\text{-}63)$$

$$\left.\begin{aligned} D_{11} &= U_1 V_{0D} + U_2 V_{1D} + U_3 V_{3D} & D_{16} &= +.5U_2 V_{2D} + U_3 V_{4D} \\ D_{22} &= U_1 V_{0D} - U_2 V_{1D} + U_3 V_{3D} & D_{26} &= +.5U_2 V_{2D} - U_3 V_{4D} \\ D_{12} &= U_4 V_{0D} - U_3 V_{3D} \\ D_{66} &= U_5 V_{0D} - U_3 V_{3D} \end{aligned}\right\} \quad (3\text{-}64)$$

where

$$V_{0D} = h^3/12$$

$$V_{1B} = \frac{1}{2}\sum_{k=1}^{n} cos(2\theta_k)(h_k^2 - h_{k-1}^2) \qquad V_{1D} = \frac{1}{3}\sum_{k=1}^{n} cos(2\theta_k)(h_k^3 - h_{k-1}^3)$$

$$\left.\begin{array}{ll}
V_{2B} = \dfrac{1}{2}\sum_{k=1}^{n} sin(2\theta_k)(h_k^2-h_{k-1}^2) & V_{2D} = \dfrac{1}{3}\sum_{k=1}^{n} sin(2\theta_k)(h_k^3-h_{k-1}^3) \\
V_{3B} = \dfrac{1}{2}\sum_{k=1}^{n} cos(4\theta_k)(h_k^2-h_{k-1}^2) & V_{3D} = \dfrac{1}{3}\sum_{k=1}^{n} cos(4\theta_k)(h_k^3-h_{k-1}^3) \\
V_{4B} = \dfrac{1}{2}\sum_{k=1}^{n} sin(4\theta_k)(h_k^2-h_{k-1}^2) & V_{4D} = \dfrac{1}{3}\sum_{k=1}^{n} sin(4\theta_k)(h_k^3-h_{k-1}^3)
\end{array}\right\} \quad (3\text{-}65)$$

The discussion above concerning the limiting values of the V_{iA} terms applies equally to the V_{iB} and V_{iD} terms; that is, the V_{iB} and V_{iD} terms, except for V_{0D}, become smaller for large variations in the θ_k. Consequently, the constant term involving V_{0D} form the limiting values of the D_{ij} terms for laminates with large numbers of different θ_k, while the limiting values of the B_{ij} are all zero. It is significant, however, that the V_{iB} and V_{iD} terms do not go to zero as fast as the corresponding V_{iA} terms, since the z coordinates occur to the second and third power (versus first power for V_{iA}), and thus the stacking sequence has a greater influence on these terms.

Example 3-6

In this example we consider an "angle-ply" laminate constructed from the nylon/elastomer material of example 2-1. An angle-ply laminate is constructed from two ply orientations which have the same magnitude but opposite signs. The laminate is balanced when there are equal numbers of plies with positive and negative orientations.

The first step is to compute the plane stress moduli:

$$Q_{11} = \frac{E_{11}}{1-\nu_{12}\nu_{21}} = \frac{132{,}000}{1-(0.36)(0.003)} = 132{,}142 \text{ psi}$$

$$Q_{22} = \frac{E_{22}}{1-\nu_{12}\nu_{21}} = \frac{1{,}050}{1-(0.36)(0.003)} = 1{,}051 \text{ psi}$$

$$Q_{12} = \nu_{21}Q_{11} = 378 \text{ psi}$$

$$Q_{66} = G_{12} = 263 \text{ psi}$$

The equivalent laminate stiffnesses are computed using: (a) equation (2-37) for the

$$\begin{aligned}
U_1 &= 50{,}173 \text{ psi} \\
U_2 &= 65{,}545 \text{ psi} \\
U_3 &= 16{,}423 \text{ psi} \\
U_4 &= 16{,}801 \text{ psi} \\
U_5 &= 16{,}686 \text{ psi}
\end{aligned}$$

(b) equation (3-62) for the A_{ij} terms; and (c) equation (3-47) for the effective moduli, $\bar{E}_{11}$.

θ	$\bar{A}_{11}$ (psi)	$\bar{A}_{22}$ (psi)	$\bar{A}_{12}$ (psi)	$\bar{E}_{11}$ (psi)
0	132,142	1,051	378	132,006
15	115,149	1,621	8,590	69,629
30	74,735	9,189	25,012	6,653
45	33,750	33,750	33,224	1,044
60	9,189	74,735	25,012	818
75	1,621	115,149	8,590	980
90	1,051	132,142	378	1,050

Figure 3-7 provides a comparison of theory and experiment for the longitudinal moduli, $\bar{E}_{11}$ for off-axis lamina (example 2-1) and angle-ply laminate. Comparison of the moduli indicates that the angle-ply laminate is much stiffer than the equivalent off-axis unidirectional lamina. The increase is caused by the constraining influence imposed on each ply within a laminate. The plies are bonded together and are not free to deform independently. If each ply had the same ν_{12} and η_{16} there would be no stiffness difference between an angle-ply and an off-axis specimen. It is for this reason that the angle-ply and off-axis properties are similar above 45°.

3.9 INFLUENCE OF THERMAL AND ENVIRONMENTAL INDUCED DILATION ON LAMINATE RESPONSE

In Section 2.4 the constitutive equation (2-38) was formulated

$$\sigma_1 = Q_{ij}\,(\varepsilon_j - e_i) \tag{2-38}$$

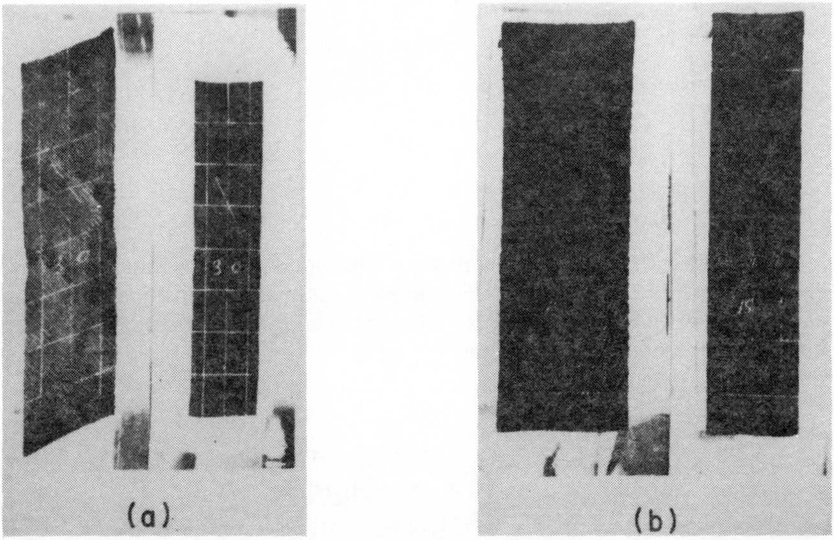

Figure 3-7. Comparison of dry and swollen 30 degree off axis lamina and a (± 15)s laminate of nylon elastomeric ply material.

which provides an interrelationship between the mechanical induces and the dilational induced expansional strain for a lamina. Following the procedures of Section 3.5 the stress resultant of equation (3-24) becomes:

$$[N] = [A][\varepsilon^\circ] + [B]\,[k] - [N]^e \tag{3-66}$$

and the moment resultants of equation (3-31) becomes:

$$[M] = [B]\,[\varepsilon^\circ] + [D]\,[k] - [M]^e \tag{3-67}$$

the new quantities N_j^e and M_j^e are defined by

$$N_j^e = \int_{-h/2}^{h/2} Q_{ij}\, e_j\, dz \tag{3-68}$$

and

$$M_j^e = \int_{-h/2}^{h/2} Q_{ij}\, e_j\, z\, dz \tag{3-69}$$

The terms N_i^e and M_i^e are called the dilational stress resultant and dilational moment resultant.

For laminates in which no bending effects are present, the strain ε in equation (3-66) are constants: hence when $N_i = 0$, we get

$$A_{ij}\, \varepsilon_j = \int_{-h/2}^{h/2} Q_{ij}\, e_j\, dz \tag{3-70}$$

where

$$A_{ij} = \int_{-h/2}^{h/2} Q_{ij}\, dz \tag{3-71}$$

Equation (3-70) defines the laminate strains under pure swelling or thermal environments and, therefore, the equivalent laminate expansional strains e_1 and e_2. The expansional strains for a balanced and symmetrical laminate can be conveniently expressed [2-6] as

$$e_1^0 = \frac{A_{22}R_1 - A_{12}R_2}{A_{11}\, A_{22} - A_{12}^2} \tag{6-23}$$

$$e_2^0 = \frac{A_{11}R_2 - A_{12}R_1}{A_{11}\, A_{22} - A_{12}^2} \tag{6-24}$$

where

$$
\begin{aligned}
R_1 &= J_1 h + J_2 H_1 = N_1^e \\
R_2 &= J_1 h - J_2 H_2 = N_2^e \\
J_1 &= (U_1 + U_4)\, W_1 + 2U_2\, W_2 \\
J_2 &= U_2\, W_1 + 2W_2\,(U_1 + 2U_3 - U_4) \\
W_1 &= \frac{1}{2}\,(e_1^\circ + e_2^\circ) \\
W_2 &= \frac{1}{4}\,(e_1^\circ + e_2^\circ) \\
e_i^\circ &= \alpha_1^\circ \Delta T + \beta_1^\circ C \\
e_z^\circ &= \alpha_2^\circ \Delta T + \beta_1^\circ C \\
H_1 &= \sum_{n=1}^{N} h_n \cos 2\theta_n \\
H_2 &= \sum_{n=1}^{N} h_n \cos 4\theta_n
\end{aligned}
\tag{3-72}
$$

and h is the laminate thickness, N the number of layers, and h_n the thickness of the nth layer. The terms R_1 and R_2 are the solutions for equation (3-68) for a balanced and symmetrical laminate. For non-symmetrical balanced laminate use equation (3-68) to evaluate $R_1 = N_1^e$ and $R_2 = N_2^e$. These expressions are fundamental to a unique property fiber reinforced composite; their ability to achieve a positive, a negative, or a zero in-plane expansional strain with thermal or moisture absorption.

Figure 3-8 illustrates a lamina and an angle ply laminate in which expansional strains were induced through the absorption of a fluid into the material (the fluid benzane into the elastomeric matrix of a tyre ply). In Figure 3-8(a) the lamina longitudinal, transverse, and shear strains are shown for a dry and swollen condition as described in Equation (2-40). Figure 3-8 (b) is the equivalent [+15/−15]s laminate which exhibits a near zero but negative longitudinal expansion despite the large positive dilation. Figure 3-8 is a comparison between theory, equations (3-26) through (3-28), and experiment for the material of examples 2-1, 2-2, and 3-6. This unique characteristic was recognized and appreciated in the mid 1960's [2-6, 3-4] and led to the first engineering applications of graphite-epoxy materials; dimensionally stable antenna dishes.

The necessity for balanced and symmetrical laminate stacking sequences can also be appreciated from the character of equations (3-66) and (3-67). For example, consider a laminate constructed of an equal number of lamina $+\theta$ and $-\theta$ orientation with all the layers of the same sign on the same side of the laminate mid plane: an antisymmetric angle-ply laminate. The resulting expansional strains and curvatures, k_j^e are:

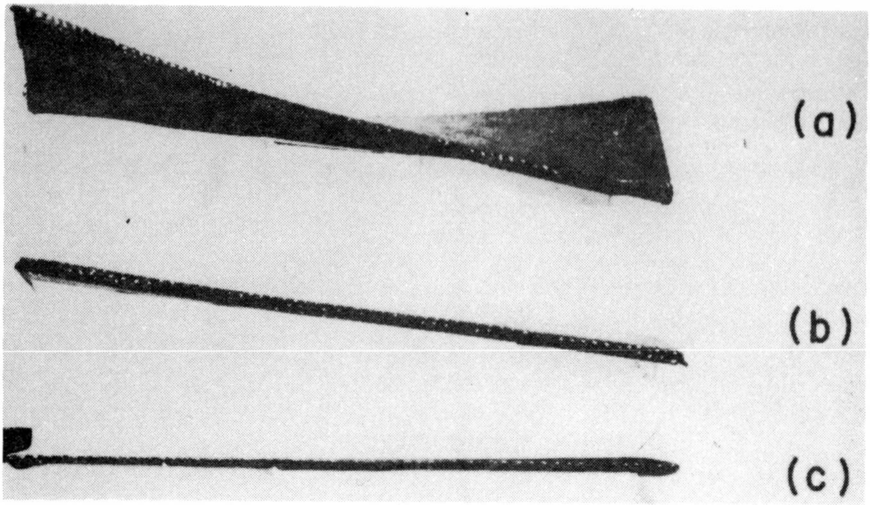

Figure 3-8. *Comparison expansional strain coupling for angle-ply laminates of a) (+ + − −), b) (+ − + −), and c) (+ − − +) fabricated from nylon elastomeric ply material.*

$$
\begin{aligned}
e_1 &= \frac{A_{22}(R_1 - B_{16}\, k_6^e) - A_{12}(R_2 - B_{26}\, k_6^e)}{A_{11}\, A_{22} - A_{12}^2} \\
e_2 &= \frac{A_{11}(R_2 - B_{26}\, k_6^e) - A_{12}(R_1 - B_{16}\, k_6^e)}{A_{11}\, A_{22} - A_{12}^2} \\
e_6 &= k_1^e = k_2^e = 0 \\
k_6^e &= M_6^e / D_{66}
\end{aligned}
\qquad (3\text{-}73)
$$

where

$$B_{16} = B_{26} = 8h^2\, U_2$$

$$D_{66} = \frac{h^3}{12}\,(U_5 - U_3 \cos 4\theta)$$

$$M_6^e = \frac{h^2}{4}\,[(Q_{11} - Q_{12})e_1 \;+\; (Q_{12} - Q_{22})e_2]\sin\, 2\theta$$

and h is the laminate thickness. Note the appearance of B_{16}, B_{26}, and the k_6^e terms in equations (3-73) versus (6-24). These terms represent a coupling between the bending and extensional deformations as illustrated in Figures 3-5 and 6-8. The nonmechanical dilational strain induces the same coupling effects as illustrated in Figure 3-8. In Figure 3-8 the laminate labeled (a) is [+15/+15/−15/−15]; in (b) [+15/−15/+15/−15]; and (c) [+15/−15/−15/+15]. The dilational strain in Figure 3-8 was induced through the volumetric contraction of the matrix as a result of cure shrinkage and the subsequent cooling from the cure condition. Only the balanced and symmetrical laminate can sustain a volumetric strain without an induced out of plane deformation.

Classical laminated plate theory predicts the shapes of all unsymmetric laminates respond to a volumetric change to be a saddle; a double deflected surface. This observation is generally true except when the length of sides of the rectilinear shape are large with respect to the thickness of the laminate. In these thin laminates, significant out of plane deflections occur resulting in geometric nonlinearities. The consequence is that as the edge length increases with respect to a constant laminate thickness the saddle configuration will disappear and two stable cylindrical configurations result. This size effect, governing the transition from stable single-valued saddle solutions to stable cylindrical solutions was developed by Hyer [3-5].

3.10 NOMENCLATURE FOR LAMINATED COMPOSITES

A laminated composite is constructed of a number of plies of unidirectional or woven fabric composites stacked at various angles relative to the x axis of the laminate. In order to describe the stacking geometry of a laminated composite, it is necessary to have a code which describes a laminate uniquely. For the case of equal ply thickness, the stacking sequence can be described by simply listing the ply orientations, θ, from top to bottom. Thus, the notation $[0/90/0]_T$ uniquely describes a three layer laminate. The subscript T denotes that the sequence accounts for the total number of layers. If a laminate is symmetric, as in the case $[0/90_2/0]_T$, the notation can be abbreviated by using $[0/90]_s$. The subscript s denotes that the stacking sequence is repeated symmetrically about the laminate centerline. The subscript *2* in total stacking sequence notation is used to show that the 90 degree ply is repeated. Angle-ply laminate are denoted by [0/45/−45], which can be shortened to $[0/\pm 45]_s$. For laminates with repeating sets of plies such as $[0/\pm 45/0/\pm 45]_s$, a shorter notation take the form $[0/\pm 45]_{2s}$. If a layer is split at the centerline in a symmetric laminate, a bar is used over the center ply to denote the split. For example, the laminate $[0/90/0]_T$ can be abbreviated to $[0/\overline{90}]_s$.

NOTATION, CHAPTER 3

$\sigma_1, \sigma_2, \sigma\tau_{12}$ = stresses in natural coordinate system
$\varepsilon_1, \varepsilon_2, \gamma_{12}$ = strains in natural coordinate system
$\sigma_x, \sigma_y, \tau_{xy}$ = stresses in arbitrary coordinate system
$\varepsilon_x, \varepsilon_y, \gamma_{xy}$ = strains in arbitrary coordinate system
$[Q]$ = stiffness matrix for a lamina in natural coordinate system
$[\bar{Q}]$ = stiffness matrix for a lamina in arbitrary coordinate system
u, v, w = displacements in the x, y, z directions
$\varepsilon_x^0, \varepsilon_y^0, \gamma_{xy}^0$ = strains at the laminate geometrical midplane
k_x, k_y, k_{xy} = curvatures of the laminate

N_x, N_y, N_{xy} = stress resultants
M_x, M_y, M_{xy} = moment resultants
$[A]$ = inplane stiffness matrix for a laminate
$[D]$ = bending stiffness matrix for a laminate
$[B]$ = coupling stiffness matrix for a laminate
h_k = layer coordinate with respect to laminate midplane
θ = orientation of natural coordinate system with respect to general coordinate system
$[A^*], [B^*], [C^*], [D^*]$ = matrices in partially inverted form of general constitutive equations
$[A'], [B'], [C'], [D']$ = matrices in totally inverted form of general constitutive equations
u_i = invariant material constants for a lamina
V_{iA}, V_{iB}, V_{iD} = laminate geometrical constants
$E_{11}, E_{22}, \nu_{12}, G_{12}$ = orthotropic elastic constants for a lamina
e_1, e_2, e_6 = expansional strains

REFERENCES

1. Assi, V. D. and Tsai, S. W., "Elastic Moduli of Laminated Anisotropic Composites," *Experimental Mechanics, 5,* 177–185 (June 1965).
2. Pister, K. S., Dong, S. B., Taylor, R. L., and Matthison, R. B., "Analysis of Structural Laminates," ARL-76 (1961).
3. Tsai, S. W. and Pagano, N. J., "Invariant Properties of Composite Materials," in *Composite Materials Workshop,* S. W. Tsai, J. C. Halpin, and N. J. Pagano, eds. Lancaster, PA: Technomic Publishing Company (1968).
4. Fahmy, A. A. and Ragai, A. N., "Thermal Expansion of Graphite-Epoxy Composites," *J. Apply. Phys., 41,* 5112 (1970).
5. Hyer, M. W., "Some Observations on the Cured Shape of Thin Unsymmetric Laminates," *J. Comp. Mat., 15,* 175 (1981); *op cit, 15,* 296 (1981).

4
Strength of Laminated Composites

4.1 INTRODUCTION

THIS CHAPTER PRESENTS A DESCRIPTION OF THE APPROACHES TO DETERmine the stress-strain response up to ultimate laminate failure for a laminated composite consisting of orthotropic lamina. The strength of a laminate will be discussed in terms of stress or strain distribution within each lamina of the laminate as constrained by a strength or ultimate strain criteria for the lamina. Laminate static strength is also influenced by the presence of residual stresses, elastic stress concentrations (notches) and interlaminar stresses which manifest themselves at the free edges of a laminate. While the concepts of fracture and delamination will be introduced, they will not be developed in detail. A detailed discussion of these issues including environmental effects and fatigue can be found in the references given at the end of this chapter.

4.2 STRENGTH OF THE ORTHOTROPIC LAMINA

The fundamental building block of composite structures is the laminate concept. In this concept the fundamental principle is that the properties of the laminate are determined by the properties of the individual layers, the lamina. In turn, it has been assumed and verified in practice that the physical properties of the lamina are unaltered by the lamination process. The strength of a laminate must necessarily be related to the individual lamina. The simplest form of implementing a failure criteria is to evaluate the failure properties of a lamina in its principle material axis for a combination of induced stresses or strains at the boundaries of the lamina.

This approach was dictated historically. It was found to be more efficient to determine where nonlinearity and degradation begin to occur on a unidirectional test specimen (lamina test) than on some general laminate test. The development of strength allowables in terms of the principle material directions was directed by the historical perception that the transformation of stress or strain is known in a linear system but the rotation of a failure surface is complex and less certain. For these reasons, the practice is to determine failure allowables for the orthotropic lamina, in simple states of stress or strain, and then to utilize analytical methods [4.1] to predict the ultimate strength of the laminate under combined loads. Historically, the strength concepts for laminated composites evolved from the classical concepts of yield theories and/or surfaces. In the discussion that follows the term "yield" will be used interchangeably with the term "strength."

Prior to any discussion of yield theories or yield surfaces for the lamina, the basic difference between the surfaces for an isotropic and for an orthotropic or anisotropic material must be explained. For an isotropic material, any biaxial stress state, σ_x, σ_y, τ_{xy} may be resolved into two principal stresses, σ_{1_p} and σ_{2_p} and some angle, θ. Since the material is isotropic, the material properties do not change with the angle θ; therefore, a plot of the principal stresses, σ_{1_p} and σ_{2_p}, which cause yield will give the required yield surface. The resulting surface is a two dimensional figure with σ_{1_p} and σ_{2_p} as axes. When considering the yield surface for an orthotropic lamina, the stresses must be referred to the lamina principal axes; therefore, for biaxial stress states, three stress components may appear in the yield criteria. The resulting yield surface will appear as a three dimensional figure with σ_1, σ_2, and τ_{12} as the reference coordinates; such a yield surface is depicted in Figure 4-1.

If the inflection point is defined as the onset of inelastic action, it is apparent that the prediction of yield strength of an orthotropic lamina is a linear problem. Several yield theories of failure have been hypothesized for anisotropic materials, and subsequently adapted to composite materials.

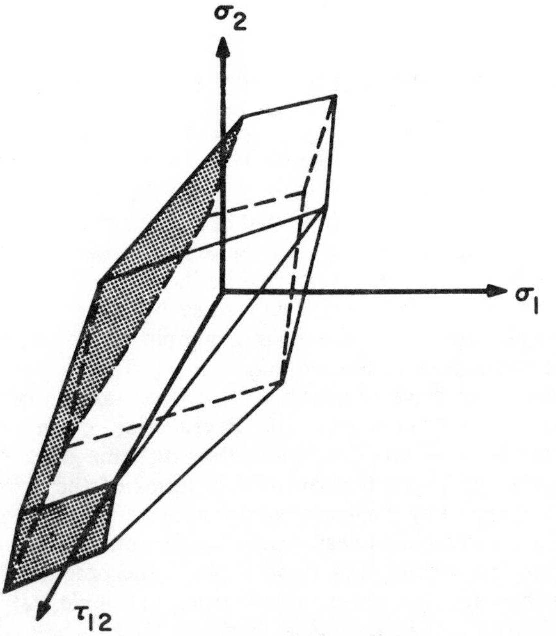

Figure 4-1. Lamina yield surface.

The yield theories which have been given the most utilization for advanced composites are the Hill theory (distortional energy) and the maximum strain theory (St. Venant). These two yield criteria will be discussed in the following two sections.

4.2.1 Hill Criterion

The Hill criterion [4-2] was developed as a generalization of the von Mises (distortional energy) isotropic yield criterion for anisotropic materials. Tsai adapted this criterion as a yield and failure criterion for laminated composites [4-3]. The development presented below is Tsai's adaptation of the Hill criterion.

For the yield strength analysis of laminated composites, the failure criterion must be based on the strengths of the individual orthotropic lamina referred to the lamina principal axes since the yield strengths are established experimentally with reference to these axes. The Hill criterion reduces to the following for an orthotropic material in plane stress (normalized with respect to the principal strengths):

$$\left(\frac{\sigma_1}{\sigma_{1_y}}\right)^2 - \frac{1}{r}\frac{\sigma_1}{\sigma_{1_y}}\frac{\sigma_2}{\sigma_{2_y}} + \left(\frac{\sigma_2}{\sigma_{2_y}}\right)^2 + \left(\frac{\tau_{12}}{\tau_{12_y}}\right)^2 = 1 \qquad (4\text{-}1)$$

where

$$\frac{1}{r} = \frac{\sigma_{1_y}}{\sigma_{2_y}}$$

Also, σ_{1_y} and σ_{2_y} are the tensile or compressive yield strengths in the 1 and 2 directions for the orthotropic lamina and τ_{12_y} is the shear yield stress.

The yield surface represented by equation (4-1) is either an ellipsoid or sphere, depending on r, when plotted in 3 dimensional space. The surface may be examined in two dimensions by observing the trace on the surface for a particular value of (τ_{12}/τ_{12_y}). For instance, if the trace at $\tau_{12} = 0$ is observed, Figure 4-2 will result. It must be kept in mind that this is the yield surface for a single orthotropic ply of a laminate. For a laminate, a set of curves such as this would be all superimposed on a set of axes which would be referenced to the laminate reference axes. That is, the axes would be σ_x, σ_y, and τ_{xy}.

The work of Hill was extended into a generalized quadratic failure criterion by Goldenblat and Kopnov [4-4]. This quadratic failure criterion has been explored in detail by Tsai and Wu [4-5].

4.2.2 Maximum Strain Criterion

The maximum strain criterion which will be presented should not be confused with the maximum principal strain criterion for isotropic materials. For the orthotropic lamina, the strain components must be referred to the lamina

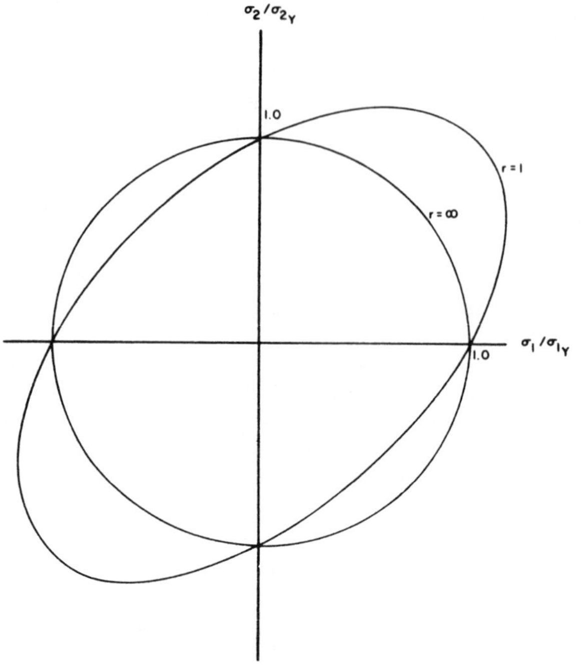

Figure 4-2. *Hill yield surfaces.*

principal axes; therefore, the possibility exists for three strain components to appear in the yield criterion.

The maximum strain criterion may be developed from equation (2-30) with the strains equal to the failure strains:

$$\begin{bmatrix} \varepsilon_{1_y} \\ \varepsilon_{2_y} \\ \gamma_{12_y} \end{bmatrix} = \begin{bmatrix} S_{11} & S_{12} & 0 \\ S_{12} & S_{22} & 0 \\ 0 & 0 & S_{66} \end{bmatrix} \begin{bmatrix} \sigma_1 \\ \sigma_2 \\ \tau_{12} \end{bmatrix} \tag{4-2}$$

Thus, equation (4-2) gives the envelope of stresses which produce the yield strains in the lamina. Failure results when any of the strain components is equal to its corresponding intrinsic strength property. Expressed in equation form, the maximum strain criterion is given as follows:

$$\begin{gathered} \varepsilon_1 \geqslant +\ \varepsilon_1,\ \varepsilon_1 > 0;\ \varepsilon_1 \geqslant -\varepsilon_1,\ \varepsilon_1 < 0 \\ \varepsilon_2 \geqslant +\ \varepsilon_2,\ \varepsilon_2 > 0;\ \varepsilon_2 \geqslant -\varepsilon_2,\ \varepsilon_2 < 0 \\ \gamma_{12} \geqslant \pm\ \varepsilon_6 \end{gathered} \tag{4-3}$$

The five intrinsic ultimate strains are defined as follows:

$+\varepsilon_1$ = Ultimate tensile strain in the fiber, (1), direction
$-\varepsilon_1$ = Ultimate compression strain in the fiber, (1), direction
$+\varepsilon_2$ = Ultimate tensile strain in the transverse, (2), direction
$-\varepsilon_2$ = Ultimate compression strain in the transverse, (2), direction
$\pm\varepsilon_6$ = Ultimate inplane, (1-2), shear strain assumed independent of the sign of the shear stress.

Five stress-strain curves are utilized because the stress-strain curves in compression differ significantly from those in tension (See Section 7.6 B & C).

4.3 LAMINATE STRENGTH ANALYSIS

Up to this point, only the elastic properties of a laminate have been considered. However, the mathematical model developed for modeling the elastic behavior of a laminate can also be used to analyze the strength behavior of a laminate. The means of performing a laminate strength analysis described below depends upon the important assumption that the behavior of each lamina in an arbitrary laminate, for given stresses (or strains) in the lamina natural axis system, is the same as the behavior measured in that natural axis system when the lamina is part of any other laminate and under the same stresses (or strains). That is, we assume that a basic strength criteria for the lamina under a state of plane stress exists, and we presume this criteria is valid for any orientation of the lamina in a laminate. The margin of strength of a laminate under a given system of loads or strains can then be determined by 1) determining the stresses and/or strains in the natural axis system for each lamina, and 2) determining the margin of each of the laminas. This procedure can be used up to the loading at which some lamina exceeds its strength. At this point, the remaining laminae must take on an additional load and may or may not fail. The proper assumption concerning what proportion of the load a failed lamina might still carry or might transfer to the other laminae is discussed in [4-1].

In the general case of an unsymmetrical laminate, the midplane strains and plate curvatures are determined using equation (3-56) once the stress resultants and moments at the point are obtained by analysis of the total structure. With these strains and curvatures known, the strain in any laminae is determined, for the x-y or laminate axis system, through equations (3-15). For a given lamina, at a given distance z from the geometrical midplane, these strains can then be transformed to the natural axis system of this lamina (1-2 system) with equations (2-35). If a strength criteria exists in terms of these natural axis system strains (such as the maximum strain theory) then a comparison can be made to determine whether the given set of loads is acceptable in terms of this criteria. If a strength criteria exists in terms of the natural axis system stresses, then these stresses can be obtained from the lamina constitutive equations (3-1), and then these stresses compared to the strength theory.

The above approach can be applied to a general laminate under general loadings (stress and moment resultants). The case when only applied stress resultants are present is of special significance. This case, which occurs in practice when laminates are used as sandwich skins, is termed a plane problem, or a membrane application. Due to manufacturing considerations these applications normally involve midplane symmetric laminates, and only such laminates are considered below.[2]

In this membrane problem, the stress resultants acting on a laminate are determined through an analysis of the total structure. Then, since the stress resultants are known and the moments are zero, equations (3-56) for $[B'] = [C'] = 0$ (midplane symmetric laminates) yield

$$\left.\begin{aligned} [\varepsilon^0] &= [A'][N] = [A^*][N] = [A^{-1}][N] \\ [k] &= [D'][M] = [0] \end{aligned}\right\} \tag{4-4}$$

Combining equations (3-61) with equation (3-14) we determine that the strains in any lamina in the x-y coordinate system are the same as the midplane strains:

$$[\varepsilon] = \begin{bmatrix} \varepsilon_x \\ \varepsilon_y \\ \gamma_{xy} \end{bmatrix} = [\varepsilon^0] + z[k] = [A^{-1}][N] + z[0] = [A^{-1}][N] \tag{4-5}$$

The strains in the natural axis system of each lamina can now be found by successively applying the transformation equations (2-35) to each lamina. If the stresses in the natural axis system are desired, they can then be determined with the lamina constitutive equation (3-1).

Example 4-1

To illustrate the point stress computations described above, we consider a three ply laminate: $+45/0/+45$. The properties and orientations are assumed to be the same as for the laminate considered in example (3-3), where it was determined that

$$[A] = 10^6 \begin{bmatrix} 7.95 & 1.75 & 1.35 \\ 1.75 & 2.55 & 1.35 \\ 1.35 & 1.35 & 1.75 \end{bmatrix}$$

We consider the case $N_x = 50{,}000$ pounds/inch, $N_y = 10{,}000$ pounds/inch, $N_{xy} = 0$. First $[A^{-1}]$ is determined to be the following (see Appendix A):

$$[A^{-1}] = 10^{-6} \begin{bmatrix} .152 & -.071 & -.062 \\ -.071 & .696 & -.482 \\ -.062 & -.482 & .991 \end{bmatrix}$$

[2]Actually, for zero curvatures (sandwich skin applications for example), the remainder of this discussion applies for unsymmetrical laminates also (although moment resultants will exist), since from equation (3-53) it is clear that the strains $[\varepsilon^0]$ are again given by $[A^*]\,[N] = [A^{-1}]\,[N]$.

Then equation (3-62) can be written

$$\begin{bmatrix} \varepsilon_x \\ \varepsilon_y \\ \gamma_{xy} \end{bmatrix} = 10^{-6} \begin{bmatrix} .152 & -.071 & -.062 \\ -.071 & .696 & -.482 \\ -.062 & -.482 & .991 \end{bmatrix} \begin{bmatrix} 50{,}000 \\ 10{,}000 \\ 0. \end{bmatrix}$$

from which

$$\begin{bmatrix} \varepsilon_x \\ \varepsilon_y \\ \gamma_{xy} \end{bmatrix} = \begin{bmatrix} .00689 \\ .00341 \\ -.00792 \end{bmatrix}$$

The stresses in the x-y coordinate system can now be determined for each lamina from the constitutive equation (3-2) for that lamina expressed in this x-y coordinate system:

$$\begin{bmatrix} \sigma_x \\ \sigma_y \\ \tau_{xy} \end{bmatrix}_{+45} = 10^6 \begin{bmatrix} 9.75 & 7.75 & 6.75 \\ 7.75 & 9.75 & 6.75 \\ 6.75 & 6.75 & 7.75 \end{bmatrix} \begin{bmatrix} -.00689 \\ .00341 \\ -.00792 \end{bmatrix} = 10^4 \begin{bmatrix} 4.01 \\ 3.31 \\ .81 \end{bmatrix}_{+45}$$

$$\begin{bmatrix} \sigma_x \\ \sigma_y \\ \tau_{xy} \end{bmatrix}_0 = \begin{bmatrix} \sigma_1 \\ \sigma_2 \\ \tau_{12} \end{bmatrix}_0 = 10^6 \begin{bmatrix} 30. & 1. & 0. \\ 1. & 3. & 0. \\ 0. & 0. & 1. \end{bmatrix} \begin{bmatrix} .00689 \\ .00341 \\ -.00792 \end{bmatrix} = 10^4 \begin{bmatrix} 21.0 \\ 1.71 \\ -.79 \end{bmatrix}_0$$

The lamina stresses and strains in the laminate x-y reference system are presented graphically in Figure 4-3.

Now that the lamina strains (or stresses) are known in the x-y system, the lamina strains (or stresses) in the natural coordinate system can be determined using the transformation equations (2-35):

$$\begin{bmatrix} \varepsilon_1 \\ \varepsilon_2 \\ \gamma_{12} \end{bmatrix}_{+45} = \begin{bmatrix} .5 & .5 & .5 \\ .5 & .5 & -.5 \\ -1. & 1. & .0 \end{bmatrix} \begin{bmatrix} .00689 \\ .00341 \\ -.00792 \end{bmatrix} = \begin{bmatrix} .00119 \\ .00911 \\ -.00348 \end{bmatrix}_{+45}$$

$$\begin{bmatrix} \sigma_1 \\ \sigma_2 \\ \tau_{12} \end{bmatrix}_{+45} = 10^4 \begin{bmatrix} .5 & .5 & 1. \\ .5 & .5 & -1. \\ -.5 & .5 & 0. \end{bmatrix} \begin{bmatrix} 4.01 \\ 3.31 \\ .81 \end{bmatrix} = 10^4 \begin{bmatrix} 4.47 \\ 2.85 \\ -.35 \end{bmatrix}_{+45}$$

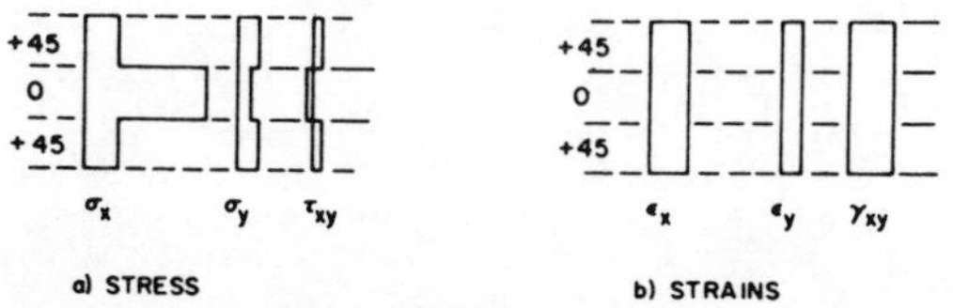

Figure 4-3. Lamina stresses and strains in x-y coordinate system.

The natural axis system for the 0-degree lamina coincides with the laminate axis system and thus the stresses and strains do not need to be transformed for this case. The stresses and strains in the natural axis system for each lamina are presented in Figure 4-4.

In the computations above, the lamina stresses (strains) in the 1-2 system were obtained from the lamina stresses (strains) in the *x-y* system using the transformation equations. An alternate procedure would be to obtain the lamina strains (stresses) in the natural axis system as above, and then use the lamina constitutive equation in the natural system, equation (3-1), to determine the corresponding stresses (strains).

Now that the lamina stresses and strains are known in the natural axis system, a strength theory for each lamina can be invoked to determine whether the loading can be carried by this laminate. For example, if the maximum strain theory is assumed, with the inequalities

$$|\varepsilon_1| \leqslant .004$$
$$|\varepsilon_2| \leqslant .003$$
$$|\gamma_{12}| \leqslant .01$$

then we find that the maximum allowable strain in the "1" direction (.004) is exceeded in the laminate under this loading in the 0-degree lamina, and that the maximum allowable strain in the "2" direction (.003) is exceeded in this laminate under this loading in all of the laminae. Thus, if the laminate is to carry this load, within the strain allowables, additional layers must be added, or a different combination of orientations must be selected. If the same orientations and relative thicknesses are to be maintained, then the total thickness must be increased by the factor

$$\frac{.00911}{.003} = 3.05$$

since the "2" direction in the +45 laminae exceeds the allowables by the greatest amount.

If a strength theory is used based upon the state of stress in the laminae, then the stresses determined in the 1-2 system for each lamina would be compared to the allowables, and the acceptability of the laminate under the given loads could again be judged.

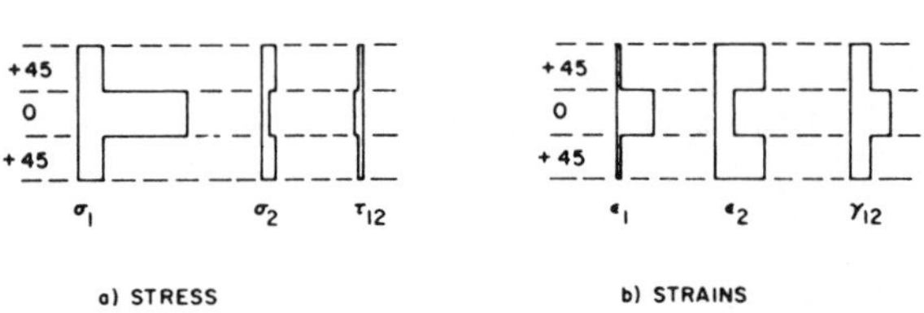

Figure 4-4. Lamina stresses and strains in 1-2 coordinate systems.

An interaction or allowable diagram for a given laminate can be determined using repetitive applications of the above procedures. To do this, various combinations of N_x, N_y, and N_{xy} are considered. If these loads are maintained at fixed ratios, then the laminate can be analyzed for the loads

$$\begin{bmatrix} N_x \\ N_y \\ N_{xy} \end{bmatrix} = \alpha \begin{bmatrix} 1 \\ P_1 \\ P_2 \end{bmatrix} \tag{4-6}$$

where

$$P_1 = \frac{N_y}{N_x} = \frac{N_y}{\alpha}$$

$$P_2 = \frac{N_{xy}}{N_x} = \frac{N_{xy}}{\alpha}$$

If the maximum values of α are determined for many ratios P_1 and P_2, then an interaction diagram in the three dimensional space with axes N_x, N_y, N_{xy} can be prepared which presents graphically the allowable stresses that the particular laminate can carry. In practice it is convenient to keep the ratio of thicknesses of each orientation in a laminate constant, but to allow the total thickness to vary. In order to prepare interaction diagrams for such laminates, the stress resultants can be divided by the thickness, thus giving average laminate stresses in the x-y coordinate system. That is, the average laminate stresses are defined as follows:

$$\begin{bmatrix} \bar{\sigma}_x \\ \bar{\sigma}_y \\ \bar{\tau}_{xy} \end{bmatrix} = \frac{1}{h} \begin{bmatrix} N_x \\ N_y \\ N_{xy} \end{bmatrix} \tag{4-7}$$

As long as the proportion of laminae at each orientation is constant, *average* allowable stresses can be determined for the laminates and presented graphically in the $\bar{\sigma}_x$, $\bar{\sigma}_y$, $\bar{\tau}_{xy}$ space. It is important to emphasize here that the stresses, $\bar{\sigma}_x$, $\bar{\sigma}_y$, $\bar{\tau}_{xy}$, are the averages of the lamina stresses, and are not the actual lamina stresses in the x-y coordinate system.

Since a three-dimensional surface is difficult to present directly, the interaction diagrams for specific laminates are generally presented in the two-dimensional $\bar{\sigma}_x$, $\bar{\sigma}_y$ system. The effect of $\bar{\tau}_{xy}$ is introduced by means of contours which show the reduction in allowable combinations of $\bar{\sigma}_x$, $\bar{\sigma}_y$ for given values of $\bar{\tau}_{xy}$. An example of such an interaction diagram for a 60 percent 0-degree, 20 percent +45 degree, 20 percent −45 degree laminate with the properties

$$E_{11} = 30 \times 10^6 \qquad G_{12} = 1.1 \times 10^6$$
$$E_{22} = 3 \times 10^6 \qquad \nu_{12} = 38$$

is given in Figure 4-5. The maximum strain theory has been used to prepare this diagram with the following allowables:

$$-.0064 \leqslant \varepsilon_1 \leqslant .0040$$
$$-.0036 \leqslant \varepsilon_2 \leqslant .0027$$
$$-.011 \leqslant \gamma_{12} \leqslant .011$$

The above discussion addressed the issue of a design allowable strength surface utilizing strain allowables less than the ultimate strain capacity of the fiber direction. In the succeeding discussion the prediction of the stress-strain response of a laminated composite up to ultimate laminate failure will be illustrated. The approach of Petit and Waddoups [4-1] will be illustrated in superficial form utilizing maximum strain as the failure criteria.

To implement maximum strain theory, the engineering moduli and ultimate strain allowables are experimentally determined for the orthotropic ply.

E_{11}—longitudinal modulus;
E_{22}—transverse modulus;
ν_{12}—shear modulus;
$\pm\varepsilon_1$—tensile and compressive longitudinal ultimate strain allowable;
$\pm\varepsilon_2$—tensile and compressive transverse ultimate strain allowable;
$\pm\varepsilon_6$—positive and negative ultimate shear strain allowable.

The plane stress moduli are computed, transformed to the plate axis system, and tabulated for each ply of a different orientation in the laminate.

The plane stress moduli are summed through the n layers of the thickness to obtain the overall laminate stiffness matrix. The stiffness matrix is inverted to obtain the overall engineering moduli of the laminate.

As the laminate is incrementally loaded, each of the plies is examined for failure. The ply failure criterion compares the actual lamina strains against

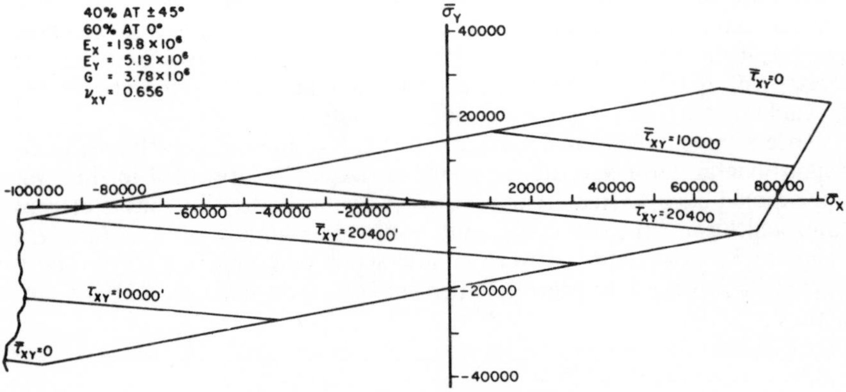

Figure 4-5. Laminate interaction diagram.

the ultimate strain allowables. If any one of the actual lamina strains exceeds the allowable, ply failure has occurred.

When the induced lamina strain equals or exceeds the lamina strain allowable, the lamina decreases in stiffness with increasing strain. It is generally assumed that any mechanical degradation (cracks) occurring at the lamina allowable strain is restricted to the damaged lamina and is not transmitted to adjacent lamina. The mechanical degradation of a lamina merely reduces the lamina modulus in the directions being considered (longitudinal, transverse, or shear). Figure 4-6 illustrates a transverse failure mode: microcracking or "crazing" induced by the transverse loads on the 90° and 45° laminas. The extent of modulus loss is proportional to the frequency of micro-cracks which increase rapidly as the transverse strain allowable is exceeded. As the incremental loads are increased, sufficient cracking will occur such that the modulus in the transverse direction will approach zero. At that time the lamina will have unloaded until virtually no load remains in the lamina in the transverse direction.

When a lamina loses its transverse load carrying capability, it is assumed that it can carry loads only parallel to the fibers or in shear. For a lamina that reaches ultimate strain in shear, it is assumed that it may still carry load parallel or transverse to the fibers. There is no ambiguity for ultimate failure in the fiber direction because this failure results in complete lamina degradation. Assumptions also must be made regarding the lamina Poisson's ratio after failure has occurred in a lamina. The minor Poisson's ratio for an orthotropic lamina can be calculated from the reciprocality relation:

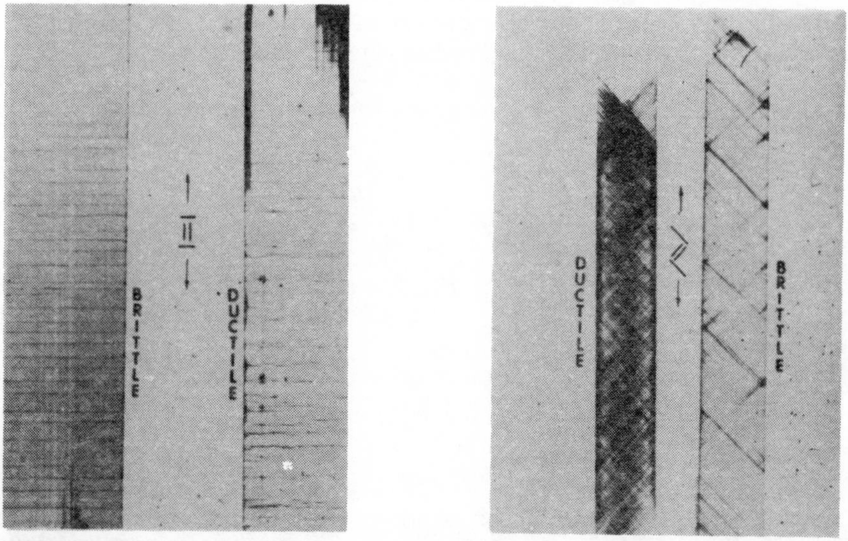

Figure 4-6. *Examples of transverse resin cracking or crazing using (0/90) and (±45) angle ply glass epoxy material [7-7].*

$$\nu_{21}E_{11} = \nu_{12}E_{22} \tag{4-8}$$

Therefore, as long as the major Poisson's ratio and the transverse modulus are defined, ν_{21} will have a value. As the modulus E_{22} decreased with increasing strain above the lamina strain allowable, E_{22} will be negative; consequently, ν_{21} will become negative. After the damaged ply completely unloads, the coupling of the Poisson's ratio is terminated and Q_{12} will become zero in the lamina stiffness matrix.

The change in the incrementally applied strain in the laminate is used to compute an incremental change in total stress using the overall laminate engineering properties. The ply direction moduli which have failed are deleted from the laminate and the calculation continues by returning to the step for calculating the overall stiffnesses. After all the plies have failed, the incremental strains and stresses are summed to obtain the ultimate stress and strain of the laminate.

A simplified version of the analytical approach of Petit and Waddoups [4-1] is illustrated in example 4-2, and the result is compared against experimental data in Figure 4-7.

Example 4-2

To illustrate the analytical procedures for estimating a laminate composite stress-strain curve, a [0/ ±45/90]s laminate fabricated from a scotch ply 1002 glass-epoxy material system is utilized. The illustration is in three steps:

Laminate Stiffness

The engineering properties characteristic of an orthotropic ply are used to determine the plane stress moduli of laminate plate theory.

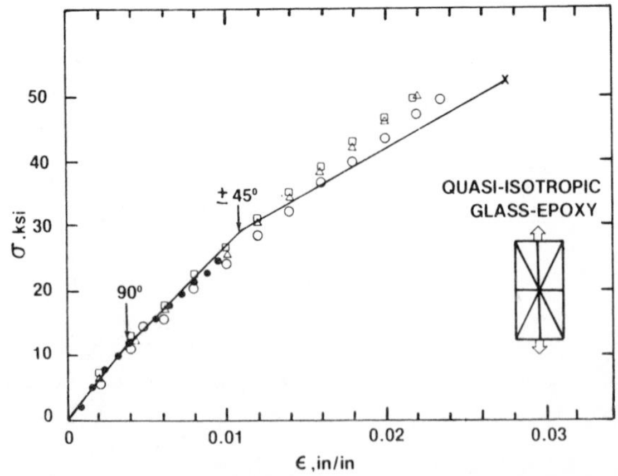

Figure 4-7. Comparison of theory and experiment for the stress-strain response of a (0/ ±45/90)s glass-epoxy laminate.

Engineering Properties

$$
\begin{aligned}
V_r &= 0.50 \\
E_{11} &= 5.6 \times 10^6 psi \\
E_{22} &= 1.2 \times 10^6 psi \\
G_{12} &= 0.6 \times 10^6 psi \\
\nu_{12} &= 0.26 \\
\nu_{21} &= \nu_{12}E_{22}/E_{11} = 0.0557
\end{aligned}
$$

Plane Stress Moduli

$$
\begin{aligned}
Q_{11} &= E_{11}/(1-\nu_{12}\nu_{21}) \\
&= 5.69 \times 10^6 psi \\
Q_{22} &= E_{22}/(1-\nu_{12}\nu_{21}) \\
&= 1.22 \times 10^6 psi \\
Q_{12} &= \nu_{21}Q_{11} = 0.317 \times 10^6 psi \\
Q_{16} &= 0 \\
Q_{26} &= 0 \\
Q_{66} &= G_{12} = 0.6 \times 10^6 psi
\end{aligned}
$$

The plane stress moduli for each ply must be transformed to the orientation the ply has in the laminate.

For the plies of the laminate

	0°	+45°	-45°	90°
$\bar{Q}_{11}$	5.68×10^6	2.48×10^6	2.48×10^6	1.22×10^6
$\bar{Q}_{22}$	1.22×10^6	2.48×10^6	2.48×10^6	5.68×10^6
$\bar{Q}_{12}$	0.316×10^6	1.28×10^6	1.28×10^6	0.316×10^6
$\bar{Q}_{16}$	0	1.12×10^6	-1.12×10^6	0
$\bar{Q}_{26}$	0	1.12×10^6	-1.12×10^6	0
$\bar{Q}_{66}$	0.6×10^6	1.57×10^6	1.57×10^6	0.6×10^6

The stiffness of the laminate is obtained by summing the plane stress moduli through the thickness in proportion to the percentage of the thickness the *k*th ply occupies of the *n* ply laminate.

$$\bar{A}_{ij} = \sum_{k-1}^{n} Q_{ij}^k a^k$$

where $a^k = \Delta h^k/h$, Δh^k = ply thickness, and h = laminate thickness.

The overall engineering properties of the laminate are obtained from the laminate stiffnesses.

$$\bar{E} = \frac{\bar{A}_{11}\bar{A}_{22} - \bar{A}_{12}^2}{A_{22}}$$

$$\bar{G} = \bar{A}_{66}$$

$$\bar{\nu} = \bar{E}\,\frac{\bar{A}_{12}}{\bar{A}_{11}\bar{A}_{22} - \bar{A}_{12}^2}$$

The overall engineering laminate properties for the quasi-isotropic laminate can be computed as the plies fail. As will be shown in the next section, the order of ply failures is 90°, ±45°, and 0°.

Laminate Stiffness as Plies Fail

Initial Laminate:

$$\begin{aligned}
\bar{A}_{11} &= \tfrac{1}{4}\,\bar{Q}_{11}^{0°} + \tfrac{1}{4}\,\bar{Q}_{11}^{+45°} + \tfrac{1}{4}\,\bar{Q}_{11}^{-45°} + \tfrac{1}{4}\,\bar{Q}_{11}^{90°} \\
&= 2.96 \times 10^6 \\
\bar{A}_{22} &= \tfrac{1}{4}\,\bar{Q}_{22}^{0°} + \tfrac{1}{4}\,\bar{Q}_{22}^{+45°} + \tfrac{1}{4}\,\bar{Q}_{22}^{-45°} + \tfrac{1}{4}\,\bar{Q}_{22}^{90°} \\
&= 2.96 \times 10^6 \\
\bar{A}_{12} &= \tfrac{1}{4}\,\bar{Q}_{12}^{0°} + \tfrac{1}{4}\,\bar{Q}_{12}^{+45°} + \tfrac{1}{4}\,\bar{Q}_{12}^{-45°} + \tfrac{1}{4}\,\bar{Q}_{12}^{90°} \\
&= 0.8 \times 10^6 \\
\bar{A}_{66} &= \tfrac{1}{4}\,\bar{Q}_{66}^{0°} + \tfrac{1}{4}\,\bar{Q}_{66}^{+45°} + \tfrac{1}{4}\,\bar{Q}_{66}^{-45°} + \tfrac{1}{4}\,\bar{Q}_{66}^{90°} \\
&= 1.08 \times 10^6 \\
\bar{E} &= 2.74 \times 10^6 \\
\bar{\nu} &= 0.27 \\
\bar{G} &= 1.08 \times 10^6
\end{aligned}$$

After the 90°ply fails:

$$\begin{aligned}
\bar{A}_{11} &= \tfrac{1}{4}\,\bar{Q}_{11}^{0°} + \tfrac{1}{4}\,\bar{Q}_{11}^{+45°} + \tfrac{1}{4}\,\bar{Q}_{11}^{-45°} \\
&= 2.66 \times 10^6 \\
\bar{A}_{22} &= \tfrac{1}{4}\,\bar{Q}_{22}^{0°} + \tfrac{1}{4}\,\bar{Q}_{22}^{+45°} + \tfrac{1}{4}\,\bar{Q}_{22}^{-45°} \\
&= 1.54 \times 10^6 \\
\bar{A}_{12} &= \tfrac{1}{4}\,\bar{Q}_{12}^{0°} + \tfrac{1}{4}\,\bar{Q}_{12}^{+45°} + \tfrac{1}{4}\,\bar{Q}_{12}^{-45°} \\
&= 0.721 \times 10^6 \\
\bar{A}_{66} &= \tfrac{1}{4}\,\bar{Q}_{66}^{0°} + \tfrac{1}{4}\,\bar{Q}_{66}^{+45°} + \tfrac{1}{4}\,\bar{Q}_{66}^{-45°} \\
&= 0.931 \times 10^6 \\
\bar{E} &= 2.32 \times 10^6 \\
\bar{\nu} &= 0.467 \\
\bar{G} &= 0.931 \times 10^6
\end{aligned}$$

After the $\pm 45°$ fails:

$$\bar{A}_{11} = \tfrac{1}{4}\,\bar{Q}_{11}^{0°} = 1.42 \times 10^6$$
$$\bar{A}_{22} = \tfrac{1}{4}\,\bar{Q}_{22}^{0°} = 0.302 \times 10^6$$
$$\bar{A}_{12} = \tfrac{1}{4}\,\bar{Q}_{12}^{0°} = 0.0792 \times 10^6$$
$$\bar{A}_{66} = \tfrac{1}{4}\,\bar{Q}_{66}^{0°} = 0.149 \times 10^6$$
$$\bar{E} = 1.40 \times 10^6$$
$$\bar{\nu} = 0.262$$
$$\bar{G} = 0.149 \times 10^6$$

Maximum Strain Failure Criterion

The orthotropic ply is characterized by six ultimate strain allowables. If any one of the ultimate strains is exceeded by a ply of the laminate, the ply has failed.

Ultimate Strain (glass-epoxy)

$$\varepsilon_1 = 0.0275$$
$$-\varepsilon_1 = 0.0158$$
$$\varepsilon_2 = 0.0038$$
$$-\varepsilon_2 = 0.0142$$
$$\varepsilon_6 = 0.0280$$
$$-\varepsilon_6 = 0.0280$$

With the laminate under uniaxial loading, the axial strain, ε_x, which causes failure in one of the plies can be computed by a transformation of the ultimate strains for a ply.

$$\varepsilon_x = \varepsilon_1/(\cos^2\theta - \bar{\nu}\sin^2\theta)$$
$$\varepsilon_x = \varepsilon_2/(\sin^2\theta - \bar{\nu}\cos^2\theta)$$
$$\varepsilon_x = \varepsilon_6/[-2\sin\theta\cos\theta(1 + \bar{\nu})]$$

The smallest axial strain which causes failure of any one of the plies of the laminate determines the order of ply failures. As each ply fails the laminate stiffness is recalculated to reflect the deletion of the failed ply. The ply failure strains and intermediate laminate moduli lead to a prediction of the stress-strain curve and the ultimate strength.

Ply failure strains for the laminate:

The $90°$ ply fails first by the positive transverse ultimate strain.

$$\theta = 90°$$
$$\varepsilon_x = \varepsilon_2/(\sin^2\theta - \bar{\nu}\cos^2\theta) = \varepsilon_2 = 0.0038$$

The $\pm 45°$ plys fail next by the positive transverse ultimate strain.

$$\theta = 45°$$
$$\varepsilon_x = \varepsilon_2/(\sin^2 \theta - \bar{\nu} \cos^2 \theta) = 0.0141$$

The 0° ply fails last by the positive longitudinal ultimate strain.

$$\theta = 0°$$
$$\varepsilon_x = \varepsilon_1/(\cos^2 \theta - \bar{\nu} \sin^2 \theta) = 0.0275$$

Summarized Data

Ply Failure	ε_x	$\bar{E}$	$\Delta\varepsilon_x$	$\Delta\sigma_x = \bar{E}\Delta\varepsilon_x$	$\Sigma\Delta\sigma_x$
90°	0.0038	2.74×10^6	0.0038	10280	10280
±45°	0.0141	2.32×10^6	0.0104	24010	34290
0°	0.0275	1.4×10^6	0.0134	18776	53050

4.4 STRENGTH OF NOTCHED LAMINATES

Fiber reinforced composites (in the fiber dominate direction) generally deform linearly to failure without yielding. This attribute creates a situation in which the fiber reinforced composite behaves more like a brittle material than a ductile metal in the presence of a notch or hole under static tension or compression test conditions. Theoretical inquiries into the phenomenon of the notch sensitivity of composites has sought to apply classical fracture concepts. These efforts have taken two forms: micro and/or macro mechanical representations. In the micro mechanics format, local fracture processes (debonding, matrix cracking, fiber breaking, etc.) are studied in the hope that mechanistic treatments can be developed for a non-homogeneous multi-angular laminate or molded part. The micro mechanical approaches face serious difficulties and have not matured at this time. The macro mechanical approaches are found on a simplified model of the composite and classical fracture mechanics for homogenous isotropic materials. The simplified composite model is the plane stress laminate plate theory which converts the non-homogenous laminated anisotropic solid into an anisotropic homogenous solid. Within the macroscopic approaches, two lines of activity exist: 1) a fracture mechanics argument [4-6 to 4-8]; and 2) a blending of classical fracture concepts and notch theory [4-9, 4-10].

Strength concepts generally imply that rupture results because the spatial average stress or strain exceeded some critical value (generally an emperical criterion) which characterized the mechanical stability of a solid. Such an attitude is useful if the microstructural perturbations to the local stress-strain fields inside a solid are of a small dimension with a low dispersion around the mean size; Region I of Figure 4-8(c). On the other hand, there are conditions in which discrete flaws, substantially larger than the uniform size distribution normally present, can exist in a material. Because they are discrete, usually relatively sharp, and larger than the surrounding disturbances, they induce

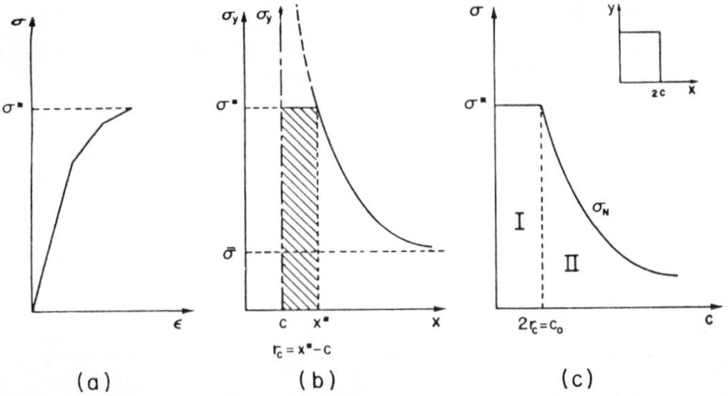

Figure 4-8. *Illustration of fracture criteria. (a) stress-strain curve for a laminate showing progressive failure of the three ply orientations, (b) condition for fracture of a material volume r_c^2 in the neighborhood of a crack-tip under uniform tensile stress in the y direction, (c) dependence of strength on flaw size.*

additional stress concentrations and provide the loci of cohesive fracture initiation. Particularly if these inherent flaws are cracklike, ordinary elastic stress-concentration factors are essentially useless because theoretical results predict an infinite concentration factor multiplying the average stress in the crack vicinity. Thus, the local stress value will exceed the finite allowable stress (or strain) experimentally measured for the base material containing only the reasonably uniform distribution of inherent flaws. The strength degradation in such a situation is illustrated by Region 11 of Figure 4-8(c). Fracture is then defined as the cleavage of a solid by a crack extension process.

4.4.1 Linear Elastic Fracture Mechanics Approach [4-6 to 4-8]

Based upon the above rationale, as illustrated in Figure 4-8, it is apparent that when one measures and defines a failure surface for a lamina (and through lamination theory for a laminate) one is actually measuring, on a spatial average basis, the criticality in stress (or strain) space of the naturally occurring flaws. Since a body containing randomly distributed microflaws is observed to sustain a set of finite loads without observable fracture, one is led to conclude that a characteristic zone of cubical dimension, r_c, must encapsulate a flaw, thus insulating the body from the infinite and hence fracture-producing stresses which would otherwise exist at the microflaw tip. In ductile metals this region, or surrounding zone, is associated with localized yielding. For a laminated material constructed of lamina with anisotropic strength characteristics, this equivalent zone is associated with localized delamination and crazing (Figure 4-6). Localized yielding or delamination and crazing result in loss of "stiffness" in the vicinity of the discontinuity, thereby limiting or "blunting" the local stresses (Figure 4-8). Fracture (that is, the rupture of a laminate in the presence of a large flaw) results when a volume element encapsulating the flaw tip experiences a local load level equal to or greater than that required to rupture the unnotched material. This critical element, characterized by the dimension r_c, serves as the link between fracture and strength analysis in solids (including composites). The hypothesis is that one need only calculate stresses outside a region r_c from a geometric discontinuity, and then view these stresses as applied to the characteristic critical zone in the same manner as occurs in the case of unflawed laminates.

This hypothesis requires that if a latent microscopic flaw is an inherent material characteristic, then the above finite volume which contains the microscopic flaw must also possess a characteristic dimension. Within this context, a measure of fracture susceptibility (or resistance) would be defined in terms of a characteristic critical zone r_c for a lamina/laminate material system. For example, Wu [4-11] and Sih [4-12] have shown that the stress σ_y (X,O) near the tip of a straing line crack of length 2C in an anisotropic plate is approximated by the expression

$$\frac{\sigma_y}{K_1} = \frac{1}{\sqrt{2\,(x\text{-}c)}} \tag{4-9}$$

where K_1 is the Mode 1 stress intensity factor defined as

$$K_1 = \bar{\sigma}\sqrt{\pi c} \tag{4-10}$$

The term σ is a uniform tensile stress applied parallel to the y axis at infinity. At fracture, the stress σ_y is equal to or less than the unnotched material strength σ^* at a distance, $X^*\text{-}C \leqslant r_c$, ahead of the crack tip (Figure 4-8(b)). This dimension represents the distance over which the material must be critically stressed in order to find a flaw sufficiently large to initiate failure.

The stress distribution ahead of a real crack of half-length c may be regarded as identical to the elastic stress (or strain) field ahead of a "notational" crack of half-length $(C + r_c)$:

$$\frac{\sigma_y}{K_1} \rightarrow \frac{\sigma^*}{K_1} \; at \; \frac{1}{\sqrt{\pi r_c}} = \frac{1}{\sqrt{\pi C_0}} \tag{4-11}$$

and

$$K_{1c} \rightarrow \sigma_n\sqrt{\pi\,(c+r_c)} = \sigma_n\sqrt{\pi\,(C+C_0)} \tag{4-12}$$

where $C_0 = r_c$. Note that as the intentionally implanted flaw of half length C goes to zero, the apparent notched strength, σ_n, goes to the pristine strength, σ^*, as defined as σ_0.

$$\sigma_n\,(c\rightarrow 0) \rightarrow \sigma_0\;\;(C<C_0) = \frac{K_{1c}}{\sqrt{\pi C_0}} \tag{4-13}$$

The apparent stress concentration K_t, will then be given as

$$\frac{\sigma_0}{\sigma_n} = K_t = \sqrt{\frac{C_0 + C}{C_0}} \tag{4-14}$$

and the critical zone dimension will be

$$C_0 = r_c = C/[(\sigma^*/\sigma_n)^2 - 1] \tag{4-15}$$

The assumption that the critical dimension r_c is a material constant is supported by the nature of equation (4-9) which states that the geometry of the stress field ahead of a line crack in an anisotropic body is independent of the degree of anisotropy of the body or of the absolute dimension of the crack.

Similar arguments may be advanced for the estimation of the critical stress for a laminate pierced by a circular hole. The above analysis attributes the contribution of a damage zone of size C_0 to be analogous to Bowie's solution [4-13] of symmetrical cracks of dimension C emanating from a circular hole of radius r in an isotropic, homogeneous material:

$$K = \sigma_n\infty \sqrt{C_0 f}\,(C_0/r) \tag{4-16}$$

or at fracture the critical stress is

$$\sigma_n\infty = \frac{K}{\sqrt{\pi C_0}\; 8\, f(C_0/r)} \tag{4-17}$$

For a specimen with no hole

$$\sigma_0 = \sigma_n\, \infty \,\Big|\, C/r \rightarrow \infty = \frac{K_c}{\sqrt{\pi C_0}}\,(1.00)$$

Thus,

$$\sigma_0/\sigma_n\infty = f(C_0/r) \tag{4-18}$$

From the ratio of the unflawed strength to the flawed specimen strength, an estimate of $f(C_0/r)$ is made. Employing tables of $f(C/r)$ a value of C/r and then C_0 is obtained. In the above expressions, C_0 for the slit and hole should be of comparable dimension for a given laminate and K_c must be the same numerical quantity if the "fracture toughness" is a material parameter. In addition, the above statements require that the apparent stress concentration $K_T = \sigma_0/\sigma_n\infty$ is dependent upon absolute hole size varying between

$$1.0\ \frac{\sigma_0}{\sigma_n\infty}\Big|_{r\rightarrow 0} \quad to \quad \frac{\sigma_0}{\sigma_n\infty} \rightarrow 1.13\, K_T \Big|_{r\rightarrow\infty}$$

in a continuous signoidal fashion. Note that in the limit of large hole size effective static stress concentration factors as estimated by anisotropic elasticity theory are observed in composites and that a hole and a slit give comparable strength reductions at small to moderate size dimensions.

4.4.2 Strength Analysis Approach

A stress criteria for fracture of laminates containing notches or holes has been developed [4-9, 4-10] based upon considerations of the exact stress gradients adjacent to either the hole or the notch. It is assumed that failure occurs when the stress over some distance away from the discontinuity is equal to or greater than the strength of the unnotched material, σ_0. It is further assumed that this characteristic distance, $\underline{d_0}$, is a material property independent of hole or notch dimension. This dimension represents the distance over which the material must be critically stressed in order to find a sufficient flaw size to initiate failure. This dimension is analogous to the approach used for predicting the plastic zone found in metals [4-14] and comparable to the arguments advanced above. Using this criteria and the approximate stress gradient for a hole of radius R in an anisotropic plate, the ratio of the notched to the unnotched strength is

$$\frac{\sigma_n}{\sigma_0} = 2/[2 + \xi^2 + 3\xi_1^4 - (K_T-3)(5\xi_1^6 - 7\xi_1^8)] \qquad (4\text{-}19)$$

where

$$\varepsilon_1 = R/(R + d_0)$$

and K_T, the stress concentration factor, is given by

$$K_T = \frac{\sigma_n}{\sigma_0} = 1 + n^* \qquad (4\text{-}20)$$

where

$$n^* = \sqrt{\frac{2}{A_{11}} \sqrt{A_{11} A_{22} - A_{12}} + \frac{A_{11} A_{12} - A_{12}^2}{2 A_{66}}}$$

or

$$n^* = \sqrt{2\left(\sqrt{\frac{\bar{E}_{11}}{\bar{E}_{22}}} - \bar{\nu}_{12} + \frac{\bar{E}_{11}}{2\,\bar{G}_{12}}\right)} \qquad (4\text{-}21)$$

The bar over the moduli denotes effective elastic module of the orthotropic laminate and the subscript one denotes the axis parallel to the applied load.

Note that for very large holes, $\varepsilon \rightarrow 1$, the classical stress concentration result is recovered, $\sigma_n \infty / \sigma_0 = 1K_T$. For vanishingly small hole sized and the rational $\sigma_n \infty / \sigma_0 \rightarrow 1.0$ as would be expected. As the anisotropic laminate goes to quasi-isotropic conditions $\bar{E} = \bar{E}_{11} = \bar{E}_{22}$, $\bar{E} = 2(1 + \bar{\nu}_{12}) \bar{G}_{12}$. For the isotropic case $(n = 2)$ in equations (4-20) and (4-21) and yields the classical solution $K_T = 3$. Equation (4-19) reduces to

$$\sigma_n / \sigma_0 = 2/(2 + \xi_1^2 + 3\xi_1^4) \tag{4-22}$$

Similar arguments for a center cracked (notched) geometry yield a ratio of the notched to the unnotched strength of

$$\sigma / \sigma_0 = \sqrt{1 - \xi_3^2} \tag{4-23}$$

where

$$\xi_3 = C/(C + d_0)$$

The predicted crack size effect on the measured value of the fracture toughness, K_q, can be developed by noting the $K_q = \sigma_n \sqrt{\pi c}$ and utilizing equation (4-23)

$$K_q = \sigma_0 \sqrt{\pi c \, (1 - \xi_3^2)} \tag{4-24}$$

In this expression, the expected limit of $K_q = 0$ for vanishing by small crack lengths is reached, while for large crack lengths, K_q asymptomatically approaches a constant value. The asymptomatic value is

$$K_q = \sigma_0 \sqrt{2\pi \, d_0}$$

which compares to previous fracture mechanics asympote

$$K_q = \sigma_0 \sqrt{\pi \, C_0}$$

It is apparent that

$$C_0 = 2d_0 \tag{4-25}$$

—Both approaches for treating data outlined above appear to yield comparable results. In fact, the approximate results of equation (4-19) can be used to estimate the $f(C_0/r)$ function in the Bowie expression, equation (4-18).

—The utility of plane stress laminated plate theory for performing strength analysis and stress analysis outside of the critical zones C_0 or d_0 appears more than adequate for both materials development and engineering design. The known problems associated with the free-edge effects matrix crazing and material non-linearity do not compromise this capability.

—As the laminates of any given material family approach the plane stress quasi-isotropic construction, the laminates reach a constant value for strength and fracture toughness properties independent of detailed angular content. Furthermore, these limiting quasi-isotropic strength and toughness properties are comparable to metallic systems. A comparison of quasi-isotropic laminate properties for different material systems with a typical aluminum alloy is given below.

Material	*Extension Stiffness (PSI)*	*Strength (KSI)*	*Fracture Toughness (KSI$\sqrt{in}$)*	*Density (lbs/in³)*
Aluminum 2024	*10.5 × 10⁶*	*62*	*32*	*0.10*
Graphite/ Epoxy	*8.0 × 10⁶*	*68*	*40*	*.054*
Boron/ Epoxy	*11.4 × 10⁶*	*60*	*32.6*	*.072*
E-Glass/ Epoxy	*2.7 × 10⁶*	*54*	*28*	*.064*
Epoxy	*0.25 × 10⁶*	*12*	*0.75*	—

—Fiber reinforced systems, although quasi-brittle in terms of fracture response, are unique in that the fracture toughness *increases* with increasing stiffness and strain to failure or strength. In most metals the toughness *decreases* with increasing yield strength. In reinforced plastics, the general rule is that if a particular reinforcement raises the ultimate strength, it will *lower* the average strain at fracture and the toughness [4-15 to 4-18]. The situation in random, chopped fiber composites, is unclear at the present time.

—The combined fiber dominance of stiffness, strength, and fracture toughness is illustrated in Figure 4-9. In this illustration the laminate fracture toughness is controlled through the distribution of lamina of different orientations within the laminate. Note that the inplane transverse direction and the response to a load or displacement normal to the lamination plane yields the lowest strength, stiffness, and fracture toughness properties.

4.5 RESIDUAL STRESSES AND STRENGTH

In Sections 2.5 and 3.9 the concept dilational strains was introduced. The dilational strains represent the volumetric changes which result from:

(a) thermal expansion

$$\left(\frac{\Delta V}{V}\right)_T = e_x^T + 2e_y^T = \alpha_x \Delta T + 2\alpha_y \Delta T;$$

(b) cure or crystallization shrinkage, [4-19]

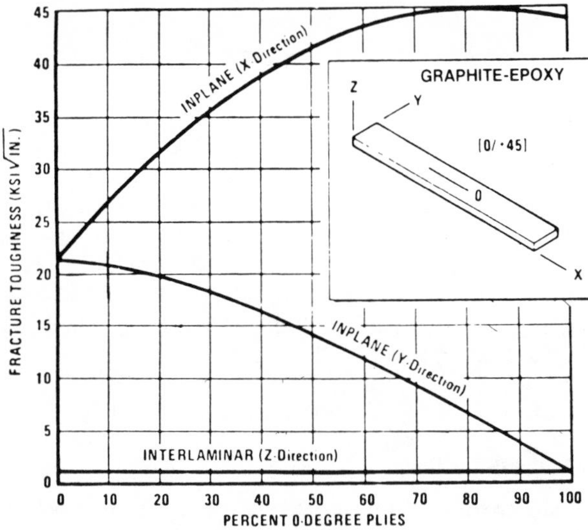

Figure 4-9. *Variation of fracture toughness with laminate content at 0 degree plus of graphite-epoxy are added to a (±45)s laminate (D. Wilkins).*

$$\left(\frac{\Delta V}{V}\right)_c = e_x^c + 2e_y^c\text{; and} \tag{4-26}$$

(c) absorption of swelling agents, such as moisture

$$\left(\frac{\Delta V}{V}\right)_s = e_x^s + 2e_y^s = \beta_x C + 2\beta_y C;$$

These expressions are an expanded form of equation (2-45). The influence of volumetric dilational changes is illustrated in Figure 4-10. These dilational strains are additive and interact with externally applied loads and/or displacements through equation (2-38) for a lamina. The equivalent stress and moment resultants are given by equations (3-66) through (3-69). For a symmetric and balanced laminate, the resulting laminate response to the volumetric changes are given by equations (3-26) and (3-27). Based on these results which establish the strain distribution in the laminate, the residual stresses in the *k*th lamina may be calculated from equation (4-27):

$$\begin{aligned}
\sigma_x^k &= [(\bar{Q}_{11}^k A_{22} - \bar{Q}_{12}^k A_{12}) R_1 \\
&\quad + (\bar{Q}_{12}^k A_{11} - \bar{Q}_{11}^k A_{12}) R_2]/(A_{11} A_{22} - A_{12}^2) \\
\sigma_y^k &= [(\bar{Q}_{12}^k A_{22} - \bar{Q}_{22}^k A_{12}) R_1 \\
&\quad + (\bar{Q}_{22}^k A_{11} - \bar{Q}_{11}^k A_{12}) R_2]/(A_{11} A_{22} - A_{12}^2)
\end{aligned} \tag{4-27}$$

$$\tau^k_{xy} = [(\bar{Q}^k_{16} A_{22} - \bar{Q}^k_{26} A_{12}) R_1 + (\bar{Q}^k_{26} A_{11} - \bar{Q}^k_{26} A_{12}) R_2]/(A_{11} - A_{22} - A^2_{12})$$

The terms R_1 and R_2 are defined in equations (3-26) and (3-27). These residual stresses may then be utilized in the stress analysis strength calculations outlined in Section 4.3 and Example 4-3.

In practice, most engineering firms have noted that failure strains, principally fiber failure strains, at the laminate level have been slightly less than characterized from unidirectional lamina. The response has been to emperically adjust and reduce the lamina allowables to be compatible with laminate experience. A more detailed discussion of the residual stresses can be found in References [4-20] and [4-22].

4.6 INTERLAMINAR STRESS AND DELAMINATION

4.6.1 General Considerations

Laminated plate theory assumes that the state of stress within each lamina of a multidirectional laminate is planar: the interlaminar stress components vanish. This assumption is accurate for inter regions removed from laminate geometric discontinuities, such as free edges. In the vicinity of the free edge a boundary layer exists where the stress is three dimensional. The boundary layer is the region wherein stress transfer between lamina is accomplished through the action of "out of plane" or laminar stresses. A simple rule of thumb [4-23] is that the boundary layer is approximately equal to the laminate thickness.

The interlaminar stresses in the boundary layer along the free edges of a laminated composite play an important role in the initiation of damage at the free edge. Delamination between layers is a direct result of these interlaminar stresses. While delamination is important to the overall structural performance of laminates under tensile loading, it is critical for compression and

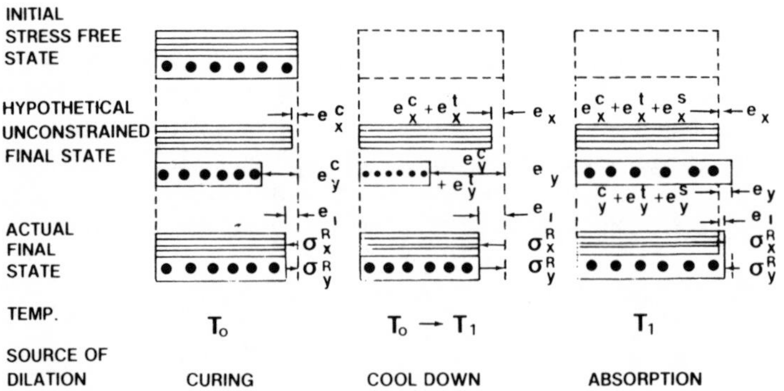

Figure 4-10. *Build up of residual stresses after curing, cooling down from cure, and absorption of a fluid such as moisture.*

shear loading where stability is a major concern. The fundamental reason for the presence of interlaminar stresses in laminated composites is the existence of a mismatch in engineering properties between layers. For the purpose of discussion, we consider here the special case of a balanced, symmetric coupon under axial loading in the x direction (Figure 4-11). The two most important lamina properties for this problem are Poisson's ratio, ν_{12} and the shear coupling term, η_{16}. Typical angular variations of ν_{12} or Q_{12} and η_{16} or Q_{16} is shown in Figures 3-6 and 2-9. The impact of the η_{16} has been shown in Figure 3-5 and is the source of the stiffness increase of the angle-ply laminate in comparison to the off-axis lamina discussed in Figure 3-6.

If there is no mismatch of ν_{12} or η_{16} between layers, there are no interlaminar stresses regardless of the mismatch in elastic and shear moduli. The mismatch of Poisson's ratio between adjacent layers give rise to dissimilar lateral strains ε_y in free (unbonded) layers, but results in identical strains at the layer interfaces with accompanying interlaminar stresses σ_y and τ_{yz} in perfectly bonded laminates. Likewise, the η_{16} mismatch gives rise to nonzero interlaminar shear stresses τ_{zx} in bonded laminates. The magnitude of the interlaminar stresses is related to the magnitude of the mismatch in ν_{xy} and η_{16}, the elastic and shear moduli and stacking sequence. The stacking sequence plays an important role in that it establishes the moment associated with the σ_y stresses. This moment is equilibrated by the interlaminar σ_x distribution (Figure 4-11).

The mechanism of interlaminar stress transfer for the $\pm\theta$ deg (angle-ply) laminate consists of laminae of only $+\theta$ deg and $-\theta$ deg fiber orientations. For a laminate subjected to axial load (x direction) only, the state of stress within each of the lamina at interior regions of the laminate is given below:

$$
\begin{aligned}
\sigma_x(\theta) &= \sigma_x(-\theta) = \overline{Q}_{11}\varepsilon_x{}^\circ + \overline{Q}_{12}\varepsilon_y{}^\circ \\
\tau_{xy}(\theta) &= -\,\tau_{xy}(-\theta) = \overline{Q}_{16}\varepsilon_x{}^\circ + \overline{Q}_{26}\varepsilon_y{}^\circ \\
\sigma_y(\theta) &= \sigma_y(-\theta) = 0
\end{aligned}
\qquad (4\text{-}28)
$$

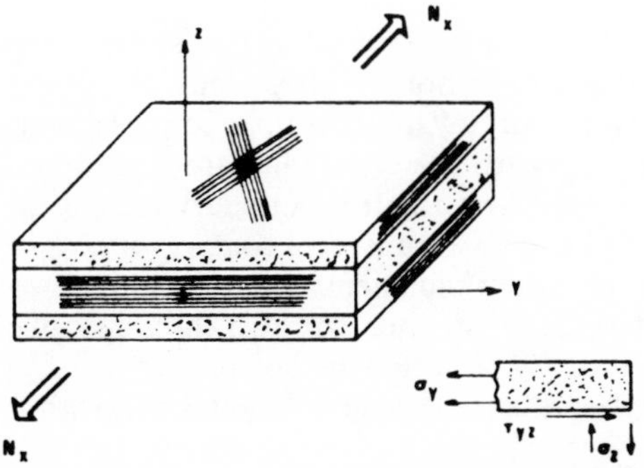

EQUILIBRIUM FREE BODY DIAGRAM

Figure 4-11. *Balanced, symmetric laminate responding to axial load.*

The inplane shear stresses within the $\pm\theta$ and $-\theta$ laminae are of equal magnitude but opposite sign. In addition, the shear stress, τ_{xy}, must vanish along free edges $y/b = 1.0$. An interlaminar shear stress τ_{xz} is required to accomplish the sign change in the shear stress τ_{xy} at the interface and to equilibrate shear stress within the laminae.

Equilibrium equations of classical theory of elasticity for a laminate of thickness h require

$$\tau_{xz}(Z) = -\int_{-h/2}^{2} \frac{\partial \tau_{xy}}{\partial_y} dn \tag{4-29}$$

For a bidirectional laminate subjected to axial load (x-direction) only, the state of stress for each lamina at interior regions of the laminate is given as:

$$\begin{aligned} \sigma_x(0) &= Q_{11}\varepsilon_x{}^{\circ} + Q_{12}\varepsilon_y{}^{\circ} \\ \sigma_x(90) &= Q_{22}\varepsilon_x{}^{\circ} + Q_{12}\varepsilon_y{}^{\circ} \\ \sigma_x(0) &= -\sigma_y(90) = Q_{12}\varepsilon_x{}^{\circ} + Q_{22}\varepsilon_y{}^{\circ} \\ \tau_{xy}(0) &= \tau_{xy}(90) = 0 \end{aligned} \tag{4-30}$$

In contrast to the state of stress for the angle-ply laminate, the inplane shear stress component vanishes for the bidirectional laminate due to the fact that the shear coupling stiffness terms $\overline{Q}_{16}$ and $\overline{Q}_{26}$ vanish at orientations of 0 deg or 90 deg. However, a mismatch between Poisson's ratios of the 0 deg and 90 deg laminae of the bidirectional laminate lead to equal but opposite-sign transverse stresses. Thus, the interlaminar stress, τ_{yz}, is required at the interface in order to accomplish the transverse stress (σ_y) sign change between the laminae. Further, the force vectors acting on the surface laminae due to the σ_y and τ_{yz} stresses are not colinear (Figure 4-11) and hence result in a couple whose magnitude is given by

$$Couple = \sigma_y h_0^2/2 \tag{4-31}$$

The couple is reacted by the interlaminar normal stress component σ_z. The distribution of the interlaminar normal stress must therefore result in zero vertical force vector while producing a couple equal in magnitude to that given by equation (4-31). When the outer plies are in transverse tension, the interlaminar normal stresses near the free boundary are tensile. The opposite is true when the outer plies are in transverse compression.

In order to demonstrate the fundamental relationship between engineering properties and the interlaminar stresses, consider three special classes of adjacent laminae: $(+\theta/-\theta)$, $(\theta/0)$, and $(\theta/90)$. The properties of greatest interest are the mismatch between adjacent layers, $\delta\nu_{12}$ and $\delta\eta_{16}$ with

$$|\delta\nu_{12}| = |\nu_{12}(\theta_2) - \nu_{12}(\theta_1)| \tag{4-32}$$

$$|\delta\eta_{16}| = |\eta_{16}(\theta_2) - \eta_{16}(\theta_1)| \tag{4-33}$$

where θ_1 and θ_2 correspond to the fiber orientations of adjacent layers. Interlaminar stresses will be largest when the mismatch is maximum (all other things being equal).

Using the graphite epoxy properties in the appendix, the following observations can be made. The largest mismatch in $|\delta\eta_{16}| = 4.34$ for the $(+11.5/-11.5)$ combination. The magnitude of $\delta\eta_{16}$ for the $(+\theta/-\theta)$ combination is exactly double that for the $(\theta/0)$ and $(\theta/90)$ combinations. This is because η_{16} is an odd function of θ and $\eta_{16} = 0$ for $\theta = 0°$ and $\theta = 90°$. There is no mismatch in ν_{12} for the $(+\theta/-\theta)$ combination as ν_{12} is an even function of θ.

The maximum value of $|\delta\nu_{12}|$ is much smaller than $|\delta\eta_{16}|$ for all three laminae combinations. The worst case for $\delta\nu_{12}$ is the $(\theta/90)$ combination with $|\delta\nu_{12}| = 0.34$ at $\theta = 22°$. Thus, the maximum value of $|\delta\eta_{16}|$ is more than ten times the maximum $|\delta\nu_{12}|$. Since the interlaminar normal and shear strengths are of the same order of magnitude, initiation of delamination is much more sensitive to the mismatch in η_{16} than the mismatch in Poisson's ratio. The zero value of $\delta\eta_{16}$ and the near maximum value of $\delta\eta_{12}$ for the (0/90) combination is consistent with the rather well-known numerical result that τ_{zx} is zero in crossply laminates and delmaination is solely due to interlaminar normal stresses.

The mismatch of the expansional strains produce edge effects similar to those due to mechanical loading (4-24). The formulation of both problems is identical and hence we consider only the thermal problem for the purpose of illustration. The mismatch in strains of interest at the free edge can be expressed

$$\delta e_y = e_y(2) - e_y(1) = (e_2 - e_1)(\cos^2\theta_2 - \cos^2\theta_1) \tag{4-34}$$
$$\delta e_{xy} = e_6(2) - e_6(2) = (e_2 - e_1)(\sin 2\theta_1 - \sin 2\theta_2) \tag{4-35}$$

using the results of equation (2-43).

—$(\pm\theta)$ Combination:

$$\begin{aligned} \delta e_y &= 0 \\ \delta e_{xy} &= (e_2 - e_1)(2\sin 2\theta) \\ |\delta e_{xy}|_{max} &= 2(e_2 - e_1) \text{ at } \theta = 45°. \end{aligned} \tag{4-36}$$

—$(\theta/0)$ Combination:

$$\begin{aligned} \delta e_y &= (e_2 - e_1)(\sin^2\theta) \\ |\delta e_y|_{max} &= (e_2 - e_1) \text{ at } \theta = 90° \\ \delta e_{xy} &= (e_2 - e_1)(\sin 2\theta) \\ |\delta e_{xy}|_{max} &= (e_2 - e_1) \text{ at } \theta = 45°. \end{aligned} \tag{4-37}$$

—(0/90) Combination:

$$\begin{aligned} \delta e_y &= (e_2 - e_1)(-\cos^2\theta) \\ |\delta e_y|_{max} &= (e_2 - e_1) \text{ at } \theta = 0° \\ |\delta e_{xy}| &= (e_2 - e_1)(\sin 2\theta) \\ |\delta e_{xy}|_{max} &= (e_2 - e_1) \text{ at } \theta = 45°. \end{aligned} \tag{4-38}$$

It is evident from the above that the worst case is δe_{xy} for the $(\pm\theta)$ combination with $\theta = 45°$. The magnitude of δe_{xy} for the $(\pm\theta)$ combination is twice that of the others, as it was for $\delta\eta_{16}$; however, the critical angle has shifted from 11.5° to 45°. The mismatch of δe_y is zero for the $(\pm\theta)$ combination, as it was for $\delta\nu_{12}$ but the maximum mismatch for the other two layer configurations both correspond to the (0/90) case unlike $\delta\nu_{12}$ where the maximum mismatch occurred at $\theta = 22°$. It is also noted that the maximum δe_{xy} is only double that of the maximum $\delta\varepsilon_y^T$ compared to a factor of more than ten between $|\delta\eta_{16}|_{max}$ and $|\delta\nu_{12}|_{max}$. Also, the variations in thermal strain mismatch as a function of θ are smooth trigonometric functions and do not exhibit the sharp rise to a peak value as does $\delta\eta_{11}$.

Since composites are cured at an elevated temperature, thermal stresses are always present. For a negative change in temperature (as during cure) the thermal shear strains e_{xy} have the opposite sign as those due to tensile loading and hence the effects tend to offset, as well as shift the critical angle. The thermal and mechanical effects would, of course, be additive for compressive loading. The combination of thermal and mechanical loading effects on $\delta\varepsilon_y$ is additive for tensile loading and offsetting for compressive loading. The absorption of a swelling agent, such as moisture, will produce effects similar to a positive temperature change.

4.6.2 Laminate Design

These results can serve as a guide to the designer in choosing fiber orientations and stacking sequence in applications with free edges. Consider the case of a laminate with fiber orientations of 0°, 90°, +45, and −45°. An often asked question is: what is the optimum stacking sequence? The curves in Figures 2-9 and 3-6 suggest that the interlaminar shear stresses τ_{zx} will be significantly lower if the +45° and −45° layers are separated by a 0° or 90° layer.

Of the 24 possible combinations of layers in the laminate, only 12 are distinct because of interchangeability of the +45° and −45° layers. Of these 12, there are six possibilities with adjacent ±45° layers and six with ±45° layers interspersed between 0° and 90° layers. These 12 laminates are depicted in Tables 4-1 and 4-2 along with the magnitude and direction of the σ_y stresses for tensile loading $\varepsilon_x = 0.5\%$. Also indicated in the tables is pertinent information as to the magnitude and direction of the equilibrating moment at each interface, and the maximum σ_x and τ_{zx} stresses determined from a finite element stress analysis.

A hierarchy of laminates for resistance to delamination can be established through consideration of the magnitude of $\delta\eta_{16}$ and the magnitude and direc-

Table 4-1. Comparative values for strength, toughness, and critical dimensions.

Laminate Construction Angles	Tensile Strength σ_0 in KSI Predicted	Experimental	Critical Dimensional, C_0 in Inches	Fracture Toughness, K_q in KSI/$\sqrt{in}$ Predicted	Experimental
T-300/5208 or AS/3501 Graphite Epoxy					
$[0]_{8S}$	190	170-190	—	95	—
$[0/90]_{4S}$	94	92	0.07-0.08	47	38
$[0/\pm 45]_{4S}$	76	76-78.5	0.07-0.08	38	30-36
$[0/\pm 60]_{4S}$	68	66	0.05	34	25
$[0/\pm 45/90]_{2S}$	68	67-69	0.06-0.08	34	26-30
$[0/\pm 36/72]_{2S}$	68	60	0.09	34	33
$[0/\pm 30/\pm 60/\text{-}90]_{2S}$	68	68	0.04	34	25
E-Glass/Epoxy (Scotch Ply 1002)					
$[0]_{4S}$	150	150	—	75	—
$[0_2/\pm 45]_{2S}$	90	87	—	45	—
$[0/90]_{2S}$	79	61-78	0.08	40	28
$[0/\pm 45/90]_{2S}$	54	46-50	0.08-0.10	25-27	22
Boron-Epoxy					
$[0]_{8S}$	192	—	—	96	—
$[0_2/\pm 45]_{2S}$	105	101	0.10-0.14	53	56-67
$[0_3/\pm 45/90]_{2S}$	—	100	0.1	—	54
$[0/\pm 45]_{2S}$	—	88	0.1	—	45
$[0/\pm 45/90]_{2S}$	70	61	0.11	35	35
Gy 70/87					
$[\pm 45/0/90]_S$			0.077		
Boron/Al					
$[\pm 45/0]_S$			0.124		

tion of the interface moment. The hierarchy given in the tables is based upon the following guidelines listed in order of severity:

1. avoid adjacent $\pm 45°$ layers
2. minimize the interface moment

The analysis indicates that the $[90/45/0/-45]_s$ laminate provides the greatest resistance to delamination for tensile loading. This laminate has the $\pm 45°$ layers interspersed and the interface moment is negative throughout. In contrast, the $[45/0/-45/90]_s$ laminate is the worst case of the six with interspersed $\pm 45°$ layers, because the interface moment is always positive and of maximum magnitude.

The hierarchy presented in the tables does not include the influence of combined stress states nor the effects of residual thermal or hygroscopic stresses. Thus, it is expected that experimental results may indicate some reordering of the list. A linear elastic finite element stress analysis of the 12 possible laminate configurations was conducted to obtain approximate elastic stress distributions. The magnitude and location of the maximum values of σ_x and

Table 4-2. Strength hierarchy for interspersed ±45° lamina.

Strength Heirarchy	Laminate Stress σ_y (KSI)	Laminate	Interface Moment $\frac{\text{in-}\ell\text{b}}{\text{in}}$	Finite Element Maximum Stresses (KSI)
1	-26.5 →	90	-0.33	
	13.5 ←	45	-0.83	σ_z = -6.8
	-0.5 →	0 ○	-1.16	τ_{zx} = -6.9
	13.5 ←	-45 □	-1.33	
2	-0.5 →	0	-0.01	
	13.5 ←	-45	0.15	σ_z = 6.2
	26.5 →	90 □ ○	0.15	τ_{zx} = 6.6
	13.5 ←	45	-0.02	
3	13.5 ←	45	0.17	
	-26.5 →	90 □	0.17	σ_z = 6.6
	-0.5 →	0 ○	-0.16	τ_{zx} = 5.9
	13.5 ←	-45	-0.33	
4	13.5 ←	45 ○	0.17	
	-26.5 →	90 □	0.17	σ_z = 6.9
	13.5 ←	-45	0.02	τ_{zx} = -6.5
	-0.5 →	0	0.02	
5	13.5 ←	45	0.17	
	0.5 →	0	0.50	σ_z = 7.6
	13.5 →	90 □ ○	0.99	τ_{zx} = -5.8
	-26.5 ←	-45	1.33	
6	13.5 ←	45	0.17	
	-0.5 →	0 ○	0.50	σ_z = 10.4
	13.5 ←	90	0.99	τ_{zx} = -6.0
	-26.5 →	-45 □	1.33	

○-maximum $|\tau_{zx}|$
□-maximum $|\sigma_z|$
T/300/5208 Graphite-Epoxy
$\varepsilon_x = 0.5\%$

τ_{zx} are presented in the tables along with a strength hierarchy based upon the following guidelines:

1. minimize $|\tau_{zx}|$
2. minimize σ_z

The finite element results clearly show the advantages to be gained with

Table 4-3. Strength hierarchy for adjacent ±45° lamina.

Strength Heirarchy	Laminate Stress σ_y (KSI)	Laminate	Interface Moment $\frac{\text{in-}\ell\text{b}}{\text{in}}$	Finite Element Maximum Stresses (KSI)
7	-26.5 →	90	-0.33	
	-0.5 →	0	-1.00	$\sigma_z = -8.2$
	13.5 ←	-45 □	-1.50	$\tau_{zx} = 9.0$
	13.5 ←	45	-1.67	
8	-26.5 →	90	-0.33	
	13.5 ←	45 □	0.82	$\sigma_z = 7.4$
	13.5 ←	-45	0.98	$\tau_{zx} = -9.2$
	-0.5 →	0	-0.98	
9	-0.5 →	0	-0.01	
	-26.5 →	90	0.34	$\sigma_z = -7.6$
	13.5 ←	45 ○	-0.85	$\tau_{zx} = -9.2$
	13.5 ←	-45 □	-1.02	
10	-0.5 →	0	-0.01	
	13.5 ←	45 ○	0.15	$\sigma_z = 10.0$
	13.5 ←	-45	0.65	$\tau_{zx} = -8.3$
	-26.5 →	90 □	0.98	
11	13.5 ←	45 ○	0.17	
	13.5 ←	-45	0.67	$\sigma_z = 9.0$
	-26.5 →	90	1.02	$\tau_{zx} = -7.7$
	-0.5 →	0 □	1.02	
12	13.5 ←	45 ○	0.17	
	13.5 ←	-45	0.67	$\sigma_z = 10.9$
	-0.5 →	0	1.34	$\tau_{zx} = -7.2$
	-26.5 →	90 □	1.67	

○-maximum $|\tau_{zx}|$
□-maximum $|\sigma_z|$
T/300/5208 Graphite-Epoxy
$\varepsilon_x = 0.5\%$

interspersed ±45° layers. All six laminates with interspersed ±45° layers have lower interlaminar shear stresses than those with adjacent ±45° layers. The magnitude of the interlaminar shear stress ranges from a low of 5.8 KSI for the $[45/0/90/-45]_s$ laminate to a high of 9.2 KSI for the $[90/45/-45/0]_s$ laminate, a 60 percent difference. The range of σ_z stresses for interspersed ±45° layers is smaller than that for the adjacent layer configuration indicating a more efficient design with interspersed ±45's in the presence of alternating positive/negative loads. The hierarchy based upon the finite element stress results is almost the same as that based upon the engineering or heuristic approach. There are a limited number of situations where the free edges are always parallel to the loading axis and the direction of loading is known to be constant. In such situations, there may be justification for adjacent $\pm\theta$ layers in order to optimize the negative interlaminar normal stress; however, this can only be accomplished at the expense of significantly higher interlaminar shear stress (Table 4-3). In many applications, the loading direction will change during service, and free edges are often present in the form of holes and cutouts eliminating the condition of a single known direction of loading. Interspersed ±0 layers are preferred in such applications as interlaminar shear stresses are reduced as are the extremes of interlaminar normal stresses.

REFERENCES

1. Petit, P. H. and Waddoups, M. E., "A Method of Predicting the Nonlinear Behavior of Laminated Composites," *J. of Comp. Mat., 3,* 2 (1969).
2. Hill, R. *The Mathematical Theory of Plasticity.* London: Oxford University Press (1950).
3. Tsai, S. W., "Strength Characteristics of Composite Materials," NASA CR-224 (April 1965).
4. Goldenblot, I. and Kopnov, V. A., "Strength of Glass-Reinforced Plastics in the Complex Stress State," *Mekhanika Polimerov, 1,* 70–78 (1965).
5. Tsai, S. W. and Wu, E. M., "A General Theory of Strength for Anisotropic Materials," *J. of Comp. Mat., 5,* 58–80 (1970).
6. Waddoups, M. E., Eisenmann, J. R., and Kaminski, B. E., *J. of Comp. Mat., 5,* 446 (1971).
7. Halpin, J. C., Jerina, K. L., and Johnston, T. A., "Analysis of the Test Methods for High Modulus Fibers and Composites," ASTM STP521, American Society for Testing and Materials, 5 (1973).
8. Waddoups, M. E. and Jalpin, J. C., *Computers and Structures, 4,* 1 (1974).
9. Whitney, J. M. and Nuismer, R. J., *J. of Comp. Mat., 8,* 253 (1974).
10. Nuismer, R. J. and Whitney, J. M., "Fracture Mechanics of Composites," ASTM STP593, American Society for Testing and Materials, 117 (1975).
11. Wu, E. M., *J. of Appl. Mech.,* Trans. ASME, Ser E, *34,* 967 (1967).
12. Sih, G. C., Paris, P. C. and Irwin, G. R., *Int. J. Fracture Mech., 1,* 189 (1964).
13. Bowie, O. L., *J. Math & Physics, 35,* 60 (1956).
14. Irwin, G. R., "Fracture Dynamics," in *Fracturing of Metals.* Cleveland: ASTM (1948).
15. Micolais, L., *Poly Engineering & Science, 15,* 137 (1975).
16. Wamback, A., Trache, K., and DiBenedetto, A., *J. of Comp. Mat., 2,* 226 (1968).
17. DiBenedetto, A. T. and Wambach, A. D., *Int. J. Polym. Material, 1,* 159 (1972).
18. Trachte, K. L. and DiBenedetto, A. T., *Int. J. Polym. Material, 1,* 75 (1971).
19. McKelvey, J. M. *Polymer Processing.* New York: John Wiley and Sons (1962).
20. Hahn, H. T. and Pagano, N. J., *J. of Comp. Mat., 9,* 91 (1975).
21. Hahn, H. T., *J. of Comp. Mat., 10,* 266 (1976).
22. Hahn, H. T. and Kim, R. Y., in *Advanced Composite Materials—Environmental Effects.* ASTM STP658, 98 (1978).
23. Pipes, R. B. and Pagano, N. J., *J. of Comp. Mat., 4,* 538 (1970).
24. Farley, G. L. and Herakovich, C. T., in *Advanced Composite Materials—Environmental Effects.* ASTM STP658 (1978).
25. Herakovich, C. T., *J. of Comp. Mat., 15,* 336 (1981).

5
Analysis of Composite Structures

5.1 INTRODUCTION

THIS CHAPTER IS DEVOTED TO A DESCRIPTION OF THE GOVERNING EQUATIONS for laminated plates and of the effects of bending-membrane coupling and anisotropy on the complexity of these equations. No attempt is made to present the many solution techniques and solutions that exist for these equations, but, rather, the simplifications introduced by special laminations and the influence of the various coupling terms are indicated. Laminated shells are not discussed in general because of the additional complexity of the governing equations and because the number of available solutions is small. The Donnell cylindrical shell equations are presented for laminated material and the presence of bending membrane coupling is noted. More thorough treatments of laminated plates and treatments of laminated shells can be found in the references at the end of this chapter.

5.2 EQUILIBRIUM EQUATIONS

The equations of equilibrium for laminated plates are identical to those applicable to plates of homogeneous material. For this reason, the equations are simply stated below. A brief derivation of the equations is presented in Appendix B.

The equations of equilibrium for a thin plate are:

$$\frac{\partial N_x}{\partial x} + \frac{\partial N_{xy}}{\partial y} = 0 \qquad (5\text{-}1)$$

$$\frac{\partial N_y}{\partial y} + \frac{\partial N_{xy}}{\partial x} = 0 \qquad (5\text{-}2)$$

$$\frac{\partial^2 M_x}{\partial x^2} + 2\frac{\partial^2 M_{xy}}{\partial x \partial y} + \frac{\partial^2 M_y}{\partial y^2} = -q(x,y) \qquad (5\text{-}3)$$

where N_x, N_y, N_{xy} are the stress resultants, M_x, M_y, M_{xy} are the moment resultant and $q(x,y)$ is the distributed transverse loading.

The first two equilibrium equations are satisfied identically by some function U, the Airy stress function, which is defined such that

$$N_x = \frac{\partial^2 U}{\partial y^2} \tag{5-4}$$

$$N_y = \frac{\partial^2 U}{\partial x^2} \tag{5-5}$$

$$N_{xy} = -\frac{\partial^2 U}{\partial x \partial y} \tag{5-6}$$

With the introduction of this stress function, equations (5-1) and (5-2) are satisfied.

In the previous chapter, we found that the constitutive equations for the moment resultants could be written, (equation (3-49)):

$$\begin{bmatrix} M_x \\ M_y \\ M_{xy} \end{bmatrix} = \begin{bmatrix} C_{11}^* & C_{12}^* & C_{16}^* \\ C_{21}^* & C_{22}^* & C_{26}^* \\ C_{61}^* & C_{62}^* & C_{66}^* \end{bmatrix} \begin{bmatrix} N_x \\ N_y \\ N_{xy} \end{bmatrix} + \begin{bmatrix} D_{11}^* & D_{12}^* & D_{16}^* \\ D_{12}^* & D_{22}^* & D_{26}^* \\ D_{16}^* & D_{26}^* & D_{66}^* \end{bmatrix} \begin{bmatrix} k_x \\ k_y \\ k_{xy} \end{bmatrix}$$

where

$[C^*] = [B][A^{-1}]$ (in general, $[C^*]$ is unsymmetric)
$[D^*] = [D] - [B][A^{-1}][B]$ (symmetric)

Introducing the Airy stress function as defined in equations (5-4) through (5-6), and the expressions for the curvatures in terms of the deflection as defined in equations (3-13), we obtain:

$$\begin{bmatrix} M_x \\ M_y \\ M_{xy} \end{bmatrix} = \begin{bmatrix} C_{11}^* & C_{12}^* & C_{16}^* \\ C_{21}^* & C_{22}^* & C_{26}^* \\ C_{61}^* & C_{62}^* & C_{66}^* \end{bmatrix} \begin{bmatrix} \dfrac{\partial^2 U}{\partial y^2} \\ \dfrac{\partial^2 U}{\partial x^2} \\ -\dfrac{\partial^2 U}{\partial y \partial y} \end{bmatrix} + \begin{bmatrix} D_{11}^* & D_{12}^* & D_{16}^* \\ D_{12}^* & D_{22}^* & D_{26}^* \\ D_{16}^* & D_{26}^* & D_{66}^* \end{bmatrix} \begin{bmatrix} -\dfrac{\partial^2 w}{\partial x^2} \\ -\dfrac{\partial^2 w}{\partial y^2} \\ -2\dfrac{\partial^2 w}{\partial x \partial y} \end{bmatrix} \tag{5-7}$$

Now these expressions for the moment resultants can be substituted into the one remaining equilibrium equation (5-3) to obtain the first governing differential equation for laminated plates:

$$\frac{\partial^2}{\partial x^2}\left(C_{11}^* \frac{\partial^2 U}{\partial y^2} + C_{12}^* \frac{\partial^2 U}{\partial x^2} - C_{16}^* \frac{\partial^2 U}{\partial x \partial y} - D_{11}^* \frac{\partial^2 w}{\partial x^2} - D_{12}^* \frac{\partial^2 w}{\partial y^2} - 2D_{16}^* \frac{\partial^2 w}{\partial x \partial y}\right)$$

$$+\,2\,\frac{\partial^2}{\partial x\partial y}\left(C_{61}^*\frac{\partial^2 U}{\partial y^2}+C_{62}^*\frac{\partial^2 U}{\partial x^2}-C_{66}^*\frac{\partial^2 U}{\partial x\partial y}-D_{16}^*\frac{\partial^2 w}{\partial x^2}\right.$$

$$\left.-D_{26}^*\frac{\partial^2 w}{\partial y^2}-2D_{66}^*\frac{\partial^2 w}{\partial x\partial y}\right)+\frac{\partial^2}{\partial y^2}\left(C_{21}^*\frac{\partial^2 U}{\partial y^2}+C_{22}^*\frac{\partial^2 U}{\partial x^2}\right.$$

$$\left.-C_{26}^*\frac{\partial^2 U}{\partial x\partial y}-D_{12}^*\frac{\partial^2 w}{\partial x^2}-D_{22}^*\frac{\partial^2 w}{\partial y^2}-2D_{26}^*\frac{\partial^2 w}{\partial x\partial y}\right)=-q(x,y)$$

or

$$C_{12}^*\frac{\partial^4 U}{\partial x^4}+(2C_{62}^*-C_{16}^*)\frac{\partial^2 U}{\partial x^3\partial y}+(C_{11}^*+C_{22}^*-C_{66}^*)\frac{\partial^4 u}{\partial x^2\partial y^2}$$

$$+\,(2C_{61}^*-C_{26}^*)\frac{\partial^4 U}{\partial x\partial y^3}+C_{21}^*\frac{\partial^4 U}{\partial y^4}$$

$$-\,D_{11}^*\frac{\partial^4 w}{\partial x^4}-4D_{16}^*\frac{\partial^4 w}{\partial x^3\partial y}$$

$$-\,2(D_{12}^*+2D_{66}^*)\frac{\partial^4 w}{\partial x^2\partial y^2}-4D_{26}^*\frac{\partial^4 w}{\partial x\partial y^3}$$

$$-\,D_{22}^*\frac{\partial^4 w}{\partial y^4}=-q(x,y) \qquad (5\text{-}8)$$

Equation (5-8) is a fourth-order partial differential equation in terms of two unknown functions, the Airy stress function U and the transverse deflection w. However, this equation contains two unknowns and cannot be solved alone. In order to solve for the deflection w and the stress function U, another equation must be found. (Note that the solution in terms of U and w will yield all the quantities of interest since knowing U we can obtain the stress resultants from equations (5-4)-(5-6), knowing the deflection w we can obtain the curvatures, and knowing the stress resultants and curvatures we can invoke the constitutive equations (3-49) to obtain the moment resultants and midplane strains.)

5.3 COMPATIBILITY CONDITION

The other governing equation needed can be found by considering the expressions for the midplane strains in terms of the midplane displacements, equations (3-13):

$$\varepsilon_x^0=\frac{\partial u_0}{\partial x}$$

$$\varepsilon_y^0 = \frac{\partial v_0}{\partial y}$$

$$\gamma_{xy}^0 = \frac{\partial u_0}{\partial y} + \frac{\partial v_0}{\partial x_0}$$

These three strain quantities, ε_x^0, ε_y^0, and γ_{xy}^0 are all given in terms of two displacements u_0, v_0, and as such cannot be taken in an arbitrary manner. In fact, differentiating ε_x^0 twice with respect to y and adding ε_y^0 differentiated twice with respect to x, we find this sum must equal γ_{xy}^0 differentiated once with respect to y and once with respect to x. That is:

$$\frac{\partial^2 \varepsilon_x^0}{\partial y^2} + \frac{\partial^2 \varepsilon_y^0}{\partial x^2} = \frac{\partial^2 \gamma_{xy}^0}{\partial x \partial y} = \frac{\partial^3 u_0}{\partial x \partial y^2} + \frac{\partial^3 v_0}{\partial x^2 \partial y} \tag{5-9}$$

Equation (5-9) is the compatibility equation for thin plates.

In the same manner as above, the midplane strains can be expressed in terms of the Airy stress function U and the deflection w by using the constitutive equations (3-49):

$$\begin{bmatrix} \varepsilon_x^0 \\ \varepsilon_y^0 \\ \gamma_{xy}^0 \end{bmatrix} = \begin{bmatrix} A_{11}^* & A_{12}^* & A_{16}^* \\ A_{12}^* & A_{22}^* & A_{26}^* \\ A_{16}^* & A_{26}^* & A_{66}^* \end{bmatrix} \begin{bmatrix} \dfrac{\partial^2 U}{\partial y^2} \\ \dfrac{\partial^2 U}{\partial x^2} \\ -\dfrac{\partial^2 U}{\partial x \partial y} \end{bmatrix} + \begin{bmatrix} B_{11}^* & B_{12}^* & B_{16}^* \\ B_{21}^* & B_{22}^* & B_{26}^* \\ B_{61}^* & B_{62}^* & B_{66}^* \end{bmatrix} \begin{bmatrix} -\dfrac{\partial^2 w}{\partial x^2} \\ -\dfrac{\partial^2 w}{\partial y^2} \\ -2\dfrac{\partial^2 w}{\partial x \partial y} \end{bmatrix} \tag{5-10}$$

where

$[A^*] = [A^{-1}]$ (symmetric)
$[B^*] = -[A^{-1}][B]$ (in general, $[B^*]$ is unsymmetric)

Now these expressions for the midplane strains can be substituted into the compatibility equation (5-9) to obtain the second governing differential equation for laminated plates:

$$\frac{\partial^2}{\partial y^2}\left(A_{11}^* \frac{\partial^2 U}{\partial y^2} + A_{12}^* \frac{\partial^2 U}{\partial x^2} - A_{16}^* \frac{\partial^2 U}{\partial x \partial y} - B_{11}^* \frac{\partial^2 w}{\partial x^2} - B_{12}^* \frac{\partial^2 w}{\partial y^2} - 2B_{16}^* \frac{\partial^2 w}{\partial x \partial y}\right)$$

$$+ \frac{\partial^2}{\partial x^2}\left(A_{12}^* \frac{\partial^2 U}{\partial y^2} + A_{22}^* \frac{\partial^2 U}{\partial x^2} - A_{26}^* \frac{\partial^2 U}{\partial x \partial y} - B_{21}^* \frac{\partial^2 w}{\partial x^2} - B_{22}^* \frac{\partial^2 w}{\partial y^2} - 2B_{26}^* \frac{\partial^2 w}{\partial x \partial y}\right)$$

$$= \frac{\partial^2}{\partial x \partial y}\left(A_{16}^* \frac{\partial^2 U}{\partial y^2} + A_{26}^* \frac{\partial^2 U}{\partial x^2} - A_{66}^* \frac{\partial^2 U}{\partial x \partial y} - B_{61}^* \frac{\partial^2 w}{\partial x^2} - B_{62}^* \frac{\partial^2 w}{\partial y^2} - 2B_{66}^* \frac{\partial^2 w}{\partial x \partial y}\right)$$

or

$$A_{22}^* \frac{\partial^4 U}{\partial x^4} - 2A_{26}^* \frac{\partial^4 U}{\partial x^3 \partial y} + (2A_{12}^* + A_{66}^*) \frac{\partial^4 U}{\partial x^2 \partial y^2} - 2A_{16}^* \frac{\partial^4 U}{\partial x \partial y} + A_{11}^* \frac{\partial^4 U}{\partial y^4}$$

$$- B_{21}^* \frac{\partial^4 w}{\partial x^4} - (2B_{26}^* - B_{61}^*) \frac{\partial^4 w}{\partial x^3 \partial y} - (B_{11}^* + B_{22}^* - B_{66}^*) \frac{\partial^4 w}{\partial x^2 \partial y^2}$$

$$- (2B_{16}^* - B_{62}^*) \frac{\partial^4 w}{\partial x \partial y^3} - B_{12}^* \frac{\partial^4 w}{\partial y^4} = 0 \qquad (5\text{-}11)$$

Equation (5-11) is a fourth order partial differential equation in terms of the two unknown functions U and w. Together with equation (5-8), we now have two such equations, so that we are, in theory, able to solve for the unknown stress function U and the deflection w subject to appropriate boundary conditions.

5.4 SIMPLIFICATION INTRODUCED BY MIDPLANE SYMMETRY

In practice, the solution of the general governing equations (5-8) and (5-11) has proven very difficult, and only a few solutions have been reported. However, just as the plate constitutive equations for the general case are simplified considerably by certain laminations, so also these simplifications carry over to the governing equations. The first such simplification is the consideration of midplane symmetric laminates. As described in Chapter 3, such midplane symmetric laminates do not exhibit coupling between bending and stretching, that is, $[B] = [0]$. As a direct result, $[B^*] = [C^*] = [0]$ and $[D^*] = [D]$.

Substituting these results in equation (5-11) we obtain an equation in terms of only the stress function U:

$$A_{11}^* \frac{\partial^4 U}{\partial y^4} - 2A_{16}^* \frac{\partial^4 U}{\partial x \partial y^3} + (2A_{12}^* + A_{66}^*) \frac{\partial^4 U}{\partial x^2 \partial y^2}$$

$$- 2A_{26}^* \frac{\partial^4 U}{\partial x^3 \partial y} + A_{22}^* \frac{\partial^4 U}{\partial x^4} = 0 \qquad (5\text{-}12)$$

This equation, with appropriate boundary conditions, governs the inplane or plane stress problem for a midplane symmetric laminated plate. The equation is identical to that governing a homogeneous (same properties through the thickness) anisotropic plate with thickness h which obeys the following constitutive law:

$$\begin{bmatrix} \sigma_x \\ \sigma_y \\ \tau_{xy} \end{bmatrix} = \frac{1}{h} \begin{bmatrix} A_{11} & A_{12} & A_{16} \\ A_{12} & A_{22} & A_{26} \\ A_{16} & A_{26} & A_{66} \end{bmatrix} \begin{bmatrix} \varepsilon_x \\ \varepsilon_y \\ \gamma_{xy} \end{bmatrix} \qquad (5\text{-}13)$$

Introducing the simplifications due to midplane symmetry into equation

(5-8), we obtain the governing differential equation for bending of such a laminate:

$$D_{11}\frac{\partial^4 w}{\partial x^4} + 4D_{16}\frac{\partial^4 w}{\partial x^3 \partial y} + 2(D_{12} + 2D_{66})\frac{\partial^4 w}{\partial x^2 \partial y^2}$$

$$+ 4D_{26}\frac{\partial^4 w}{\partial x \partial y^3} + D_{22}\frac{\partial^4 w}{\partial y^4} = q(x,y) \qquad (5\text{-}14)$$

This equation, with appropriate boundary conditions, governs the bending problem for a midplane symmetric laminated plate. The equation is identical to that governing a homogeneous anisotropic plate with thickness h which obeys the following constitutive law:

$$\begin{bmatrix} \sigma_x \\ \sigma_y \\ \tau_{xy} \end{bmatrix} = \frac{12}{h^3}\begin{bmatrix} D_{11} & D_{12} & D_{16} \\ D_{12} & D_{22} & D_{26} \\ D_{16} & D_{26} & D_{66} \end{bmatrix}\begin{bmatrix} \varepsilon_x \\ \varepsilon_y \\ \gamma_{xy} \end{bmatrix} \qquad (5\text{-}15)$$

Equations (5-12) and (5-14), which govern the inplane and bending problems for midplane symmetric laminated plates, have been found more tractable than the coupled equations (5-8) and (5-11). However, the unsymmetric differentiations $\partial^4/\partial x \partial y^3$ and $\partial^4/\partial y \partial x^3$ render these equations intractable by many of the approaches used for isotropic plates.

5.5 SPECIALLY ORTHOTROPIC LAMINATES

A further simplification is introduced if the inplane behavior is specially orthotropic, that is, if $A_{16} = A_{26} = 0$. Then the plane stress problem is governed by the equation:

$$A_{11}^* \frac{\partial^4 U}{\partial y^4} + (2A_{12}^* + A_{66}^*)\frac{\partial^4 U}{\partial x^2 \partial y^2} + A_{22}^* \frac{\partial^4 U}{\partial x^4} = 0 \qquad (5\text{-}16)$$

This equation is essentially of the same form as that governing an isotropic plate, and thus the methods used to obtain isotropic plane stress solutions are usually applicable.

Similarly, if the bending behavior is specially orthotropic, that is, if $D_{16} = D_{26} = 0$, then the bending problem is governed by an equation which is essentially of the same form as the bending equation for isotropic plates:

$$D_{11}\frac{\partial^4 w}{\partial x^4} + (2D_{12} + D_{66})\frac{\partial^4 w}{\partial x^2 \partial y^2} + D_{22}\frac{\partial^4 w}{\partial y^4} = q(x,y) \qquad (5\text{-}17)$$

This equation can usually be solved by the same methods used to obtain solutions for the bending of isotropic plates.

5.6 SIGNIFICANCE OF BENDING-MEMBRANE COUPLING

It is beyond the scope of this chapter to discuss the various techniques applicable to the solution of the governing differential equations for the bending of plates. The references listed at the end of this chapter provide a wealth of such discussion and numerous examples. However, it is within the scope of the present discussion to present a few representative solutions to illustrate the effects of the bending-membrane coupling terms and the normal moment —twist curvature coupling.

As an illustrative example, we consider a rectangular simply-supported cross-plied[3] plate under the action of a sinusoidal distributed loading

$$q(x,y) = q_0 \sin\left(\frac{\pi x}{a}\right) \sin\left(\frac{\pi y}{b}\right) \tag{5-18}$$

where a is the dimension in the x direction and b is the dimension in the y direction. The maximum deflection is sought. Equations (5-8) and (5-11) have been solved by Whitney for this case and the reader is referred to reference [5-5] for a description of the solution method. For our purposes, we list only Whitney's solution for the maximum deflection:

$$w_{max} = q_0 \frac{a^4}{\pi^4} B \tag{5-19}$$

where

$$B = \frac{\left[\left(A_{11} + A_{66}\frac{a^2}{b^2}\right)\left(A_{66} + A_{11}\frac{a^2}{b^2}\right) - (A_{12} + A_{66})^2 \frac{a^2}{b^2}\right]}{\left[\left(A_{11} + A_{66}\frac{a^2}{b^2}\right)\left(A_{66} + A_{11}\frac{a^2}{b^2}\right) - (A_{12} + A_{66})^2 \frac{a^2}{b^2}\right]\left[D_{11}\left(1 + \frac{a^4}{b^2}\right) + 2(D_{12} + 2D_{66})\frac{a^2}{b^2}\right] - B_{11}^2\left[A_{11}\left(1 + \frac{a^6}{b^6}\right) + A_{66}\left(1 + \frac{a^8}{b^8}\right) + 2(A_{12} + A_{66})\frac{a^4}{b^4}\right]}$$

Note that when $B_{11} = 0$ (when the membrane-bending coupling disappears) the solution reduces to

$$B = \frac{1}{D_{11}\left(1 + \frac{a^4}{b^4}\right) + 2(D_{12} + 2D_{66})\frac{a^2}{b^2}}$$

[3]In the present context, cross-plied implies a laminate fabricated of alternating layers of 0 and 90 degree orientations. Furthermore, we will consider only laminates with an even number of layers.

which is the solution obtained for a simply-supported orthotropic plate if the coupling is zero or is not included in the analysis.

The significance of the coupling terms can be demonstrated by considering a numerical example. We consider a plate fabricated of N pairs of cross-plied laminae with $E_{11} = 30. \times 10^6$ psi, $E_{22} = 0.75 \times 10^6$ psi, $G_{12} = .7 \times 10^6$ psi, and $\nu_{12} = .25$. Solutions for a plate with $a/b = 2$ are presented in Figure 5-1 for the maximum deflection vs N as obtained including the membrane-bending coupling, and as obtained with the orthotropic analysis. Note that the difference between the simple orthotropic solution and the correct solution is large for small values of N, but decreases rapidly as N grows large (as the bending-membrane coupling decreases). Also note that the orthotropic solution underestimates the maximum deflection.

These results are typical of those obtained to date for unsymmetrical laminates. They indicate that bending-membrane coupling cannot be ignored in either the analysis or characterization, Figures (7-5) and (7-7), of laminated structures, and that the orthotropic solution neglecting this coupling is non-conservative.

5.7 SIGNIFICANCE OF THE D_{16}, D_{26} TERMS

Most practical laminates are constructed with midplane symmetry, and thus the bending-membrane coupling is eliminated. However, as indicated in Chapter 3, the bending stiffness matrix $[D]$ is fully populated in the general case of midplane symmetric laminates and in fact the D_{16} and D_{26} terms are exactly zero for only laminates fabricated solely of 0 degree and 90 degree laminae. The general case of bending of midplane symmetric laminates is thus covered by equation (5-14). As indicated above, this equation is more difficult to solve than the less general equation (5-17) which applies to orthotropic plates. Since equation (5-17) is easier to solve, many of the solutions

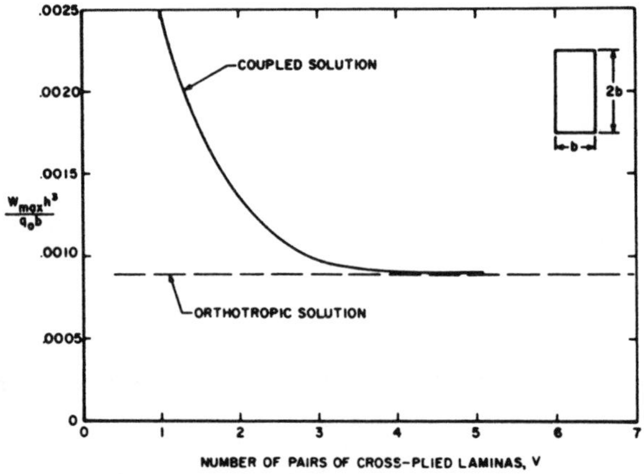

Figure 5-1. Maximum deflections of simply-supported unsymmetrical laminates.

utilized for laminated plates are based upon such orthotropic analysis. The example below is intended to illustrate the potential pitfall of such a simplification.

We consider a square plate with edges fixed to prevent rotation or translation. The plate is composed of a single orthotropic layer with the axis of orthotropy oriented at an angle θ with respect to the plate edges. The *[D]* matrix for any orientation except $\theta = 0°$ or $90°$ contains D_{16} and/or $D_{26} \neq 0$. The plate is under the action of a uniformly distributed load and has the following properties: $E_{11} = 30. \times 10^6$, $E_{22} = 3. \times 10^6$, $G_{12} = .75 \times 10^6$, $\nu_{12} = .3$. Figure 5-2 presents solutions from reference [5-10] obtained with an anisotropic analysis (including D_{16} and D_{26}) and solutions obtained ignoring the coupling terms. The anisotropic solutions for the maximum deflection are always greater than those obtained with the orthotropic analysis except at $\theta = 0°$ and $90°$. Thus, we find in this case that use of the orthotropic analysis can introduce appreciable errors into the analysis of a midplane symmetric laminated plate and these errors are in a non-conservative direction. The results indicated in the example above are representative of the results obtained to date. The effect of the anisotropic coupling terms D_{16} and D_{26} is usually such as to render an orthotropic analysis non-conservative, and thus the indiscriminate use of such orthotropic analysis is not warranted.[4]

5.8 LAMINATED BEAMS

A common geometric construction form is a beam. In this section, equations which are applicable to a general class of symmetric laminates are

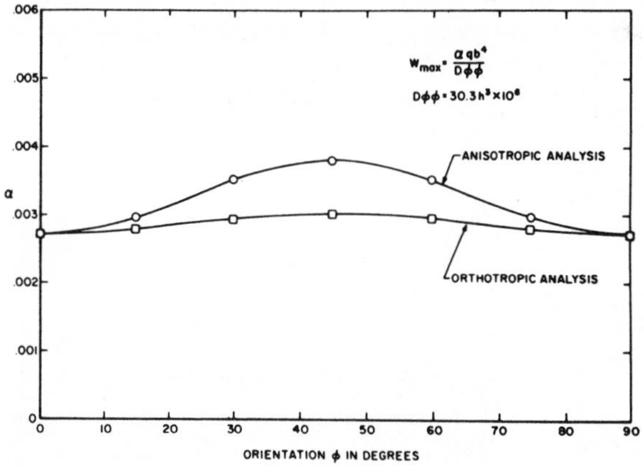

Figure 5-2. *Maximum deflection versus orientation, square clamped anisotropic plate.*

[4]It should be added that the same trends mentioned for the case of transverse loadings have been found in the stability and dynamic analysis of laminated plates.

developed for a beam as a special limit of laminated plate theory [5-12]. In order to derive a beam theory the following assumptions are made:

$$M_y = M_{xy} = 0 \tag{5-20}$$

Using equation (5-2) in conjunction with equations (3-13) and (3-52) yields the result

$$\frac{\partial^2 w}{\partial x^2} = -D_{11} M_x \tag{5-21}$$

Since beams have a high length-to-width ratio it is assumed that

$$w = w(x) \tag{5-22}$$

Caution must be exercised in applying equation (5-22) to composite beams. In particular, equations (3-52) and (5-20) imply that both the curvatures K_y and K_{xy} are functions of the bending moment M_x, that is

$$\frac{\partial^2 w}{\partial y^2} = -D'_{12} M_x \quad \frac{\partial^2 w}{\partial x \partial y} = -\frac{D'_{16}}{2} M_x \tag{5-23}$$

Thus, the deflection, w, cannot be independent of y. Even in homogeneous isotropic beam theory the one-dimensional assumption is not strictly correct due to the effect of Poisson's ratio, D'_{12} in equation (5-23). The effect is negligible, however, if the length-to-width ratio is moderately large. In the case of anisotropic shear/coupling, as displayed by D'_{16} in equation (5-22), the effect can be more severe. This is of particular importance for angle-ply laminates. The twisting curvature induced by the D'_{16} term can cause the specimen to lift off its supports at the corners. This point will be further discussed in Chapter 7.

Equation (5-22) in conjunction with equation (5-21) leads to the result

$$\frac{d^2 w}{dx^2} = \frac{M}{\bar{E}I} \tag{5-24}$$

where

$$\bar{E} = \frac{12}{h^3 D_{11}{}^*}, \quad M = bM_x$$

and b is the width of the beam. Equation (5-24) is in the same form as classical beam theory with the isotropic modulus E replaced by the effective bending modulus of the laminated beam, $\bar{E}$.

The equilibrium equation for the bending of a symmetrically laminated plate is of the form

$$\frac{\partial^2 M_x}{\partial x^2} + \frac{2\partial^2 M_{xy}}{\partial x \partial y} + \frac{\partial^2 M_y}{\partial y^2} + p = 0 \qquad (5\text{-}25)$$

where p is a normal stress distributed over the top surface of the plate. Substituting equations (3-22) and (5-25) and taking equations (5-20) and (5-22) into account yields the relationship

$$\frac{d^4 w}{dx^4} = D_{11}{}^* p \qquad (5\text{-}26)$$

Multiplying equation (5-26) by the beam width b yields

$$\frac{d^4 w}{dx^4} = \frac{q}{\overline{EI}} \qquad (5\text{-}27)$$

where

$$q = bp$$

Again, equation (5-27) is analogous to classical homogeneous, isotropic beam theory. Equilibrium considerations also yield the following relationship from laminated plate theory:

$$\frac{\partial M_x}{\partial x} + \frac{\partial M_{xy}}{\partial y} - Q_x = 0 \qquad (5\text{-}28)$$

where Q_x is the transverse shear resultant per unit width of the plate strip. Substituting equations (5-20) into (5-28) and multiplying the result by b leads to the relationship

$$Q = \frac{dM}{dx} \qquad (5\text{-}29)$$

where

$$Q = bQ_x$$

Equation (5-28) is exactly the same as in the homogeneous isotropic beam theory. Such a result is anticipated since equation (5-29) is simply a statement of equilibrium between the bending moment and transverse shear resultant.

Stresses in the kth layer of the beam are given by the relationship

$$\begin{bmatrix} \sigma_x^k \\ \sigma_y^k \\ \sigma_{xy}^k \end{bmatrix} = \begin{bmatrix} Q_{11}^k & Q_{12}^k & Q_{16}^k \\ Q_{12}^k & Q_{22}^k & Q_{26}^k \\ Q_{16}^k & Q_{26}^k & Q_{66}^k \end{bmatrix} \begin{bmatrix} \varepsilon_x \\ \varepsilon_y \\ \varepsilon_{xy} \end{bmatrix} \tag{5-30}$$

where, in the usual manner, stress is denoted by σ and engineering strains by ε.

For symmetric laminates under bending loads

$$\varepsilon_x = zK_x,\ \varepsilon_y = zK_y,\ \varepsilon_{xy} = zK_{xy} \tag{5-31}$$

where z is the coordinate through the plate thickness. Substituting equations (5-31) into (5-30), taking equation (3-52) into account, and multiplying the results by b yields

$$\sigma_x^k = z\, f_1^k \frac{M}{I} \tag{5-32}$$

$$\sigma_y^k = z\, f_2^k \frac{M}{I} \tag{5-33}$$

$$\sigma_{xy}^k = z\, f_3^k \frac{M}{I} \tag{5-34}$$

where

$$f_1^k = (Q_{11}^k S_{11b} + Q_{12}^k S_{12b} + Q_{16}^k S_{16b})$$
$$f_2^k = (Q_{12}^k S_{11b} + Q_{22}^k S_{12b} + Q_{26}^k S_{16b})$$
$$f_3^k = (Q_{16}^k S_{11b} + Q_{26}^k S_{12b} + Q_{66}^k S_{16b})$$

and

$$S_{ijb} = \frac{D_{ij}{}^* h^3}{12}$$

For homogeneous beams $f_1^k =$ unity, $f_2^k = f_3^k = 0$, and equation (5-32) reduces to classical beam theory, while equations (5-33) and (5-34) vanish. Attention should be focused on the multiaxial stresses in a laminated beam as displayed by equations (5-32) and (5-34). These equations are not correct in a zone of approximately one laminate thickness, h, away from the free-edge where state of stress is three-dimensional. This inaccuracy arises from the fact that the plate theory only requires the vanishing of resultant moments along the free edges rather than a point by point vanishing of the stresses.

The interlaminar shear stress is often of interest because of the relatively weak interlaminar shear strength encountered in laminate composite materials. The interlaminar shear stress σ_{xz}^k can be determined from the following equilibrium equation of elasticity:

$$\frac{\partial \sigma_x^k}{\partial x} + \frac{\partial \sigma_{xy}^k}{\partial y} + \frac{\partial \sigma_{xz}^k}{\partial z} = 0 \tag{5-35}$$

Assuming the stresses are independent of y, substituting equation (5-18) into (5-21), taking equation (5-29) into account, and integrating the results with respect to z yields the following equation for interlaminar shear stress:

$$\sigma_{xz}^k = -\frac{Q}{I}\int_{-h/2}^{z} f_1^k \, z dz \tag{5-36}$$

The integral assures continuity of the transverse shear stress at the layer interfaces. For homogeneous materials equation (5-36) becomes

$$\sigma_{xz} = (h^2 - 4z^2)\frac{Q}{8I} \tag{5-37}$$

which is the relationship obtained from classical beam theory.

A beam theory which includes the effect of transverse shear deformation can be derived from an existing laminated plate theory in a matter analogous to the Kirchhoff-type analysis. The results are equivalent to the Timoshenko beam theory for homogeneous isotropic materials [5-13].

Using the theory developed by Whitney and Pagano [5-14], the constitutive relations in equation (3-52), taking equation (5-20) into account, are of the form

$$\begin{bmatrix} \dfrac{\partial \psi_x}{\partial_x} \\ \dfrac{\partial \psi_y}{\partial_y} \\ \dfrac{\partial \psi_x}{\partial_y} + \dfrac{\partial \psi_y}{\partial_x} \end{bmatrix} = \begin{bmatrix} D_{ij}^* \end{bmatrix} \begin{bmatrix} M_x \\ 0 \\ 0 \end{bmatrix} \tag{5-38}$$

where ψ_x and ψ_y are rotations of the normal to the mid-plane relative to the x and y axes, respectively. The constitutive relations in shear take the form [5-14]

$$\begin{bmatrix} Q_x \\ Q_y \end{bmatrix} = k \begin{bmatrix} A_{55} & A_{45} \\ A_{45} & A_{44} \end{bmatrix} \begin{bmatrix} \psi_x + \dfrac{\partial w}{\partial x} \\ \psi_y + \dfrac{\partial w}{\partial y} \end{bmatrix} \tag{5-39}$$

where

$$A_{ij} = \int_{-h/2}^{h/2} C_{ij} dz$$

and C_{ij} are anisotropic transverse shear stiffnesses. The k factor is a parameter used to improve the results of the plate theory as compared with theory of elasticity. Equilibrium considerations in conjunction with equation (5-22) show that Q_y vanishes. Thus, the inverted form of equation (5-39) takes the form

$$\begin{bmatrix} \psi_x + \dfrac{\partial w}{\partial x} \\ \\ \psi_y + \dfrac{\partial w}{\partial y} \end{bmatrix} = \frac{1}{k} \begin{bmatrix} A_{55}^* & A_{45}^* \\ \\ A_{45}^* & A_{44}^* \end{bmatrix} \begin{bmatrix} Q_x \\ \\ 0 \end{bmatrix} \tag{5-40}$$

where

$$A_{55}^* = \frac{A_{44}}{A}, \quad A_{45}^* = -\frac{A_{45}}{A}, \quad A_{44}^* = \frac{A_{55}}{A}$$

and

$$A = (A_{44}A_{55} - A_{45}^2)$$

It is assumed that ψ_x is independent of y, that is,

$$\psi_x = \psi(x) \tag{5-41}$$

Substituting equations (5-38), (5-39), and (5-41) into equation (5-28) yields

$$\frac{d^2\psi}{dx^2} - \left(\psi + \frac{dw}{dx}\right) \frac{k\bar{G}bh}{\bar{EI}} = 0 \tag{5-42}$$

where

$$\bar{G} = \frac{1}{hA_{55}^*}$$

From equilibrium considerations in classical laminated plate theory

$$\frac{\partial Q_x}{\partial x} + \frac{\partial Q_y}{\partial y} + p = 0 \tag{5-43}$$

Substituting equations (5-40) into (5-43), taking into account equation (5-41), the vanishing of Q_y, and multiplying the results by b leads to the result

$$\frac{d^2w}{dx^2} + \frac{d\psi}{dx} + \frac{q}{k\bar{G}bh} = 0 \tag{5-44}$$

Equations (5-42) and (5-44) are of the same form as the Timoshenko beam theory [5-13]. For most cases the classical value of 5/6 can be used for k. It is usually not necessary to adjust the stresses for shear deformation. As a result, expressions for the stresses in conjunction with the shear deformation beam theory are not presented. Classical beam theory is recoverd for layered beams in which the layers of the beam are: a) stacked symmetrically about the midplane; b) are isotropic; and/or c) the orthotropic axes of the material symmetry in each ply are parallel to the beam edges. The bending stiffness EI is replaced by the equivalent stiffness $E_{11_b}I$ defined [5-15, 5-16].

$$E_{11_b}I = \sum_{k=1}^{n} E_{11}{}^{(k)}I^k \tag{5-45}$$

where E_{11_b} is the effective bending modulus of the beam, $E_{11}{}^k$ is the modulus of the kth layer relative to the beam axis, I is the moment of inertia of the beam relative to the midplane, I^k is the moment of inertia of the kth layer relative to the midplane, and n is the number of layers in the laminate. For rectangular cross sections, the bending or torsional stiffness are:

$$\frac{E\,I}{b} = \frac{1}{D_{11}{}^*}$$

$$\frac{G\,J}{b} = \frac{4}{D_{66}{}^*}$$

Note that for laminated anisotropic systems EI and GJ are not coupled as they are in isotropic materials.

5.9 LAMINATED CYLINDRICAL SHELLS

Although general laminated shells are beyond the scope of this presentation, a special class of important laminated shells will be considered in this section. Circular cylindrical shells are considered briefly; the governing equations for these shells are quite similar to the governing equations for plates, and the development is entirely analogous. Cylindrical shells are of special importance in composite material characterization, since such cylindrical specimens are often used to obtain combined stress states.

Since the development of the governing equations for cylindrical shells follows closely the development presented earlier for plates, only a brief development is included below. The references listed at the end of the chapter provide thorought development of the general theory of shells, as well as more detailed developments for cylindrical shells.

We consider a circular cylinder of radius r in cylindrical coordinates x, θ, z as shown in Figure 5-3. We will restrict ourselves to the Donell Approximations [5-18]. The strain-displacement relations can then be shown to be

$$\begin{bmatrix} \varepsilon_x \\ \varepsilon_\theta \\ \gamma_{x\theta} \end{bmatrix} = \begin{bmatrix} \varepsilon_x^0 \\ \varepsilon_\theta^0 \\ \gamma_{x\theta}^0 \end{bmatrix} + z \begin{bmatrix} k_x \\ k_\theta \\ k_{x\theta} \end{bmatrix} \tag{5-46}$$

where

$\begin{bmatrix} \varepsilon_x \\ \varepsilon_\theta \\ \gamma_{x\theta} \end{bmatrix}$ = normal and shear strains at a distance z from the geometrical mid-plane of the shell

$$\begin{bmatrix} \varepsilon_x^0 \\ \varepsilon_\theta^0 \\ \gamma_{x\theta}^0 \end{bmatrix} = \text{strains at the geometrical mid-plane} = \begin{bmatrix} \dfrac{\partial u_0}{\partial x} \\ \dfrac{1}{r}\left(\dfrac{\partial v_0}{\partial \theta} + w\right) \\ \dfrac{\partial v_0}{\partial x} + \dfrac{1}{r}\dfrac{\partial u_0}{\partial \theta} \end{bmatrix} \tag{5-47}$$

u_0, v_0, w = mid-plane displacements in the x, θ, z directions.

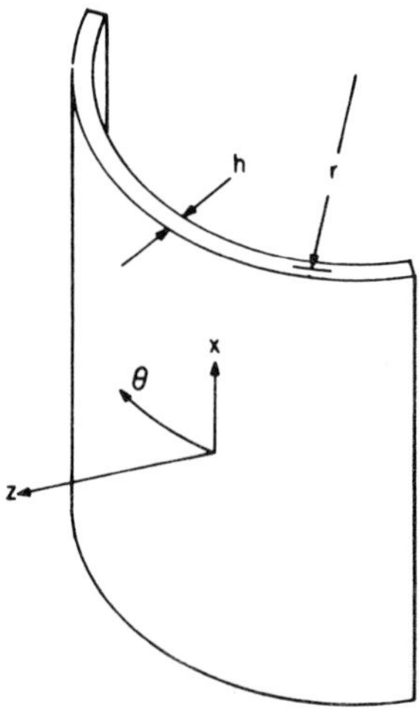

Figure 5-3. Cylindrical shell coordinates.

$$\begin{bmatrix} k_x \\ k_\theta \\ k_{x\theta} \end{bmatrix} = -\begin{bmatrix} \dfrac{\partial^2 w}{\partial x^2} \\ \dfrac{1}{r^2}\dfrac{\partial^2 w}{\partial \theta^2} \\ \dfrac{2}{r^2}\dfrac{\partial^2 w}{\partial x \partial \theta} \end{bmatrix} \tag{5-48}$$

Equation (5-46) is identical in form to equation (3-14) for plates.

For shells such that z/r can be neglected with respect to 1., the stress and moment resultants for the cylindrical shell are identical to those defined in Chapter 3. Since the general constitutive equation (3-33) was derived from the definitions of the stress and moment resultants together with the strain-displacement relations (analogous to equations (5-46)) and the constitutive relations for the layers, equations (3-2), the same process leads immediately to the constitutive equations for thin cylindrical shells:

$$\begin{bmatrix} N_x \\ N_\theta \\ N_{x\theta} \end{bmatrix} = \int_{-h/2}^{h/2} \begin{bmatrix} \sigma_x \\ \sigma_\theta \\ \tau_{x\theta} \end{bmatrix} dz$$

$$= \begin{bmatrix} A_{11} & A_{12} & A_{16} \\ A_{12} & A_{22} & A_{26} \\ A_{16} & A_{26} & A_{66} \end{bmatrix} \begin{bmatrix} \varepsilon_x^0 \\ \varepsilon_\theta^0 \\ \gamma_{x\theta}^0 \end{bmatrix} + \begin{bmatrix} B_{11} & B_{12} & B_{16} \\ B_{12} & B_{22} & B_{26} \\ B_{16} & B_{26} & B_{66} \end{bmatrix} \begin{bmatrix} k_x \\ k_\theta \\ k_{x\theta} \end{bmatrix} \tag{5-49}$$

$$\begin{bmatrix} M_x \\ M_\theta \\ M_{x\theta} \end{bmatrix} = \int_{-h/2}^{h/2} \begin{bmatrix} \sigma_x \\ \sigma_\theta \\ \tau_{x\theta} \end{bmatrix} z\, dz$$

$$= \begin{bmatrix} B_{11} & B_{12} & B_{16} \\ B_{12} & B_{22} & B_{26} \\ B_{16} & B_{26} & B_{66} \end{bmatrix} \begin{bmatrix} \varepsilon_x^0 \\ \varepsilon_\theta^0 \\ \gamma_{x\theta}^0 \end{bmatrix} + \begin{bmatrix} D_{11} & D_{12} & D_{16} \\ D_{12} & D_{22} & D_{26} \\ D_{16} & D_{26} & D_{66} \end{bmatrix} \begin{bmatrix} k_x \\ k_\theta \\ k_{x\theta} \end{bmatrix} \tag{5-50}$$

or

$$\begin{bmatrix} N \\ \hline M \end{bmatrix} = \left[\begin{array}{c|c} A & B \\ \hline B & D \end{array}\right] \begin{bmatrix} \varepsilon^0 \\ \hline k \end{bmatrix} \tag{5-51}$$

where

$$A_{ij} = \sum_{k=1}^{n} (\bar{Q}_{ij})_k \, (h_k - h_{k-1})$$

$$B_{ij} = \frac{1}{2} \sum_{k=1}^{n} (\bar{Q}_{ij})_k \, (h_k^2 - h_{k-1}^2)$$

$$D_{ij} = \frac{1}{3} \sum_{k=1}^{n} (\bar{Q}_{ij})_k \, (h_k^3 - h_{k-1}^3)$$

The general constitutive equations for thin laminated cylindrical shells thus have the same form as the general constitutive equations for laminated plates. Note that, in general, bending-membrane coupling exists due to the B_{ij} terms.

In the development of the governing partial differential equations for plates, the constitutive relation (3-33), in the partially inverted form of equation (3-49), was used in conjunction with the equations of equilibrium and a compatibility condition. This process again can be followed for cylindrical shells. The equations of equilibrium are:

$$\frac{\partial N_x}{\partial x} + \frac{1}{r} \frac{\partial N_{x\theta}}{\partial \theta} = 0 \tag{5-52}$$

$$\frac{\partial N_{x\theta}}{\partial x} + \frac{1}{r} \frac{\partial N_{\theta}}{\partial \theta} = 0 \tag{5-53}$$

$$\frac{\partial^2 M_x}{\partial x^2} + \frac{2}{r} \frac{\partial^2 M_{x\theta}}{\partial x \partial \theta} + \frac{1}{r^2} \frac{\partial^2 M_\theta}{\partial \theta^2} - \frac{N_\theta}{r} = -q(x,\theta) \tag{5-54}$$

where $q(x,\theta)$ is the distributed loading in the z direction. Equations (5-52) and (5-53) can again be satisfied identically if we utilize a stress function to replace the stress resultants, defined as follows:

$$\left. \begin{aligned} N_x &= \frac{1}{r^2} \frac{\partial^2 U}{\partial \theta^2} \\ N_\theta &= \frac{\partial^2 U}{\partial x^2} \\ N_{x\theta} &= -\frac{1}{r} \frac{\partial^2 U}{\partial x \partial \theta} \end{aligned} \right\} \tag{5-55}$$

Utilizing this stress function, the third equilibrium equation, and the partially inverted form of the general constitutive relations, equation (3-49), the first governing partial differential equation for thin cylindrical shells is found:

$$C_{12}^* \frac{\partial^4 U}{\partial x^4} + \frac{(2C_{62}^* - C_{16}^*)}{r} \frac{\partial^4 U}{\partial x^3 \partial \theta} + \frac{(C_{11}^* + C_{22}^* - 2C_{66}^*)}{r^2} \frac{\partial^4 U}{\partial x^2 \partial \theta^2}$$
$$+ \frac{(2C_{61}^* - C_{26}^*)}{r^3} \frac{\partial^4 U}{\partial x \partial \theta^3} + \frac{C_{21}^*}{r^4} \frac{\partial^4 U}{\partial \theta^4} - D_{11}^* \frac{\partial^4 w}{\partial x^4} - 4 \frac{D_{16}^*}{r} \frac{\partial^4 w}{\partial x^3 \partial \theta}$$

$$- \frac{2(D_{12}^* + 2D_{66}^*)}{r^2} \frac{\partial^4 w}{\partial x \partial \theta^2} - 4 \frac{D_{26}^*}{r^3} \frac{\partial^4 w}{\partial x \partial \theta^3} - \frac{D_{22}^*}{r^4} \frac{\partial^4 w}{\partial \theta^4} - \frac{1}{r} \frac{\partial^2 U}{\partial x^2}$$

$$= -q(x,\theta) \qquad (5\text{-}56)$$

Equation (5-56) is the same form as equation (5-8) for plates except for the factors involving the radius r, and the term

$$- \frac{1}{r} \frac{\partial^2 u}{\partial x^2}$$

This latter term involving the stress function does not disappear when the bending-stretching coupling in the constitutive equation is eliminated (i.e., when $[B] = [0]$) and thus the governing equations for shells are coupled even for midplane symmetric laminates.

The second governing equation for cylindrical shells is found, as for plates, by utilizing a compatibility condition:

$$- \frac{1}{r} \frac{\partial^2 w}{\partial x^2} + \frac{\partial^2 \varepsilon_\theta^0}{\partial x^2} + \frac{1}{r^2} \frac{\partial^2 \varepsilon_x^0}{\partial \theta^2} = \frac{1}{r} \frac{\partial^2 \gamma_{r\theta}^0}{\partial x \partial \theta} \qquad (5\text{-}57)$$

Using this equation with the definitions (5-55) and the partially inverted form of the constitutive equations (3-49), the second governing partial differential equation is obtained:

$$A_{22}^* \frac{\partial^4 U}{\partial x^4} - 2 \frac{A_{26}^*}{r} \frac{\partial^4 U}{\partial x^3 \partial \theta} + \frac{(2A_{12}^* + A_{66}^*)}{r^2} \frac{\partial^4 U}{\partial x^2 \partial \theta^2} - 2 \frac{A_{16}^*}{r^3} \frac{\partial^4 U}{\partial x \partial \theta^3}$$

$$+ \frac{A_{11}^*}{r^4} \frac{\partial^4 U}{\partial \theta^4} - B_{12}^* \frac{\partial^4 w}{\partial x^4} - \frac{(2B_{26}^* - B_{61}^*)}{r} \frac{\partial^4 w}{\partial x^3 \partial \theta}$$

$$- \frac{(B_{11}^* + B_{22}^* - 2B_{66}^*)}{r^2} \frac{\partial^4 w}{\partial x^2 \partial \theta^2} - \frac{(2B_{16}^* - B_{62}^*)}{r^3} \frac{\partial^4 w}{\partial x \partial \theta^3}$$

$$- \frac{B_{12}^*}{r^4} \frac{\partial^4 w}{\partial \theta^4} - \frac{1}{r} \frac{\partial^2 w}{\partial x^2} = 0 \qquad (5\text{-}58)$$

Equation (5-58) is analogous to equation (5-11) for plates. Note here again, however, that even when the constitutive equations are uncoupled (when $[B] = [0]$), this equation still involves both U and W and is thus coupled due to the addition of the term

$$- \frac{1}{r} \frac{\partial^2 w}{\partial x^2}$$

The constitutive equation for the mechanical response of this unidirectional and laminated cylindrical shells has been developed by Whitney and Halpin [5-20] and is expressed as

$$
\begin{aligned}
\varepsilon_x^0 &= (A_{11}^\circ + A_{16}^\circ K_1)N_x + (A_{12}^\circ + A_{16}^\circ K_2)\,Rp + A_{16}^\circ K_3 T \\
\varepsilon_y^0 &= (A_{12}^\circ + A_{26}^\circ K_1)N_x + (A_{22}^\circ + A_{26}^\circ K_2)\,Rp + A_{26}^\circ K_3 T \\
\varepsilon_{xy}^0 &= (A_{16}^\circ + A_{66}^\circ K_1)N_x + (A_{26}^\circ + A_{66}^\circ K_2)\,Rp + A_{66}^\circ K_3 T
\end{aligned}
\qquad (5\text{-}59)
$$

where

$$
A^\circ = A^{-1}
$$
$$
B^\circ = -A^{-1}B
$$

The stress resultant induced by an internal pressure p's

$$
N_y = Rp
$$

The total torque T applied to the tube is given by

$$
T = 2\pi R\,(RN_{xy} + M_{xy}) = 2\pi R\,[RN_{xy} - B_{16}^\circ N_x - B_{26}^\circ Rp]
$$

or

$$
N_{xy} = K_1 N_x + K_2 Rp + K_3 T
$$

where

$$
K_1 = \frac{B_{16}^\circ}{R},\; K_2 = \frac{B_{26}^\circ}{R},\; K_3 = \frac{1}{2\pi R^2}
$$

For symmetric layups $(B_{ij} = 0)$

$$
K_1 = K_2 = 0,\; K_3 = \frac{1}{2\pi R^2}
$$

The tensile stress N_x is given as σ_x/h where h is the thickness of the shell. In unidirectional systems, equation (5-59) is uncoupled, $B_{ij} = 0$; therefore, $K_1 = K_2 = 0$. Let us consider the response of a 45° helical tube constructed from the fiber reinforced material characterized in Figure 2-12. The result of a tensile stress $\sigma_x = hN_x$, where h is the thickness of the shell, gives the response

$$
\varepsilon_x^0 = A_{11}^\circ N_x,\; \varepsilon_y^0 = A_{12}^\circ N_x, \text{ and } \varepsilon_{xy}^0 = A_{16}^\circ N_x
$$

where $A_{ij}^\circ = A_{ij}^{-1}$ or $hA_{ij}^\circ = \bar{S}_{ij}$. $\bar{S}_{ij}$ represents the compliance moduli averaged through the thickness of the shell wall. The response to a positive N_x is then exactly equivalent, Figure 5-4B, to the tensile strip experiment performed in Figure 2-10. The angle rotation is, of course, a measure of A_{16}°.

The application of an internal pressure, p, into a right cylindrical specimen results in an $N_y = Rp$ and an $N_x = \frac{1}{2} N_y$, therefore

$$\varepsilon_x^0 = (\tfrac{1}{2} A_{11}^{\circ} + A_{12}^{\circ})\, Rp$$
$$\varepsilon_y^0 = (\tfrac{1}{2} A_{12}^{\circ} + A_{22}^{\circ})\, Rp$$
$$\varepsilon_{xy}^0 = (\tfrac{1}{2} A_{16}^{\circ} + A_{26}^{\circ})\, Rp$$

This response is shown in Figure 5-4C. The angle of rotation is now a measure of A_{16}° and A_{26}°. In Figures 5-4D and 4-4E are shown the combined influence of a tension, N_x, or compression, $-N_x$, and internal pressure, $N_x = \frac{1}{2} N_y$ plus N_y, note changing sign of N_x changes sign of the shear coupling rotation A_{16}°. Similar experiments are recorded in Figure 5-5 in which A_{16}° and A_{26}° have opposite signs: note $S_{16} = hA_{16}^{\circ}$ etc. Figure 5-5E indicates a state of biaxial tension in which N_x and N_y have been adjusted so that the $-A_{16}^{\circ}$ responses cancel out the A_{26}° rotation. The application of a torque to obtain a shear stress N_{xy} is also described by equation (5-59) in a similar fashion to yield a measure of $\bar{S}_{66}$ or A_{66}°. Note here that the application of a torque will also result in an extension or contraction in longitudinal and transverse directions due to the coupling terms A_{16}° and A_{26}°.

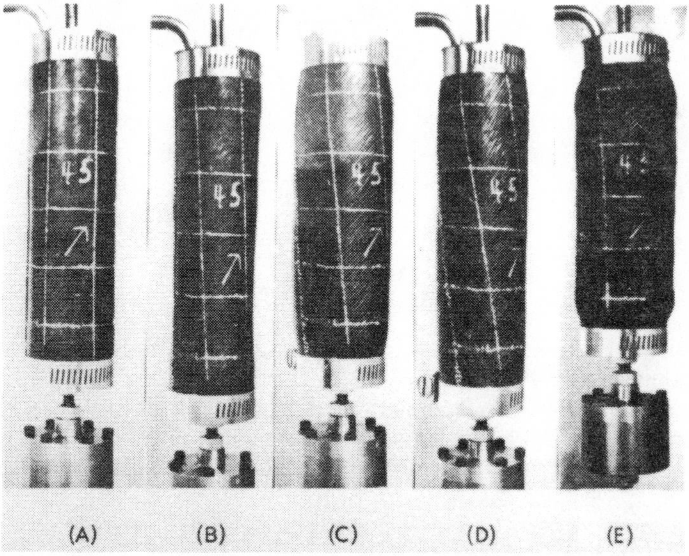

Figure 5-4. *45° cylindrical shell of nylon elastomeric ply material, S_{16} and S_{26} are positive: (A) undeformed; (B) tension, N_x; (C) internal pressure, $N_x = \frac{1}{2} N_y$, N_y; (D) tension plus internal pressure; and (E) compression plus internal pressure.*

(A) (B) (C) (D) (E)

Figure 5-5. *75° cylindrical shell of nylon elastomeric ply material,* S_{16} *negative and* S_{26} *positive: (A) undeformed; (B) internal pressure,* $N_x = \frac{1}{2} N_y$, N_y*; (C) tension,* N_x*; (D) tension plus internal pressure; and (E) tension plus decreased internal pressure.*

Let us now consider the response of a two-layer cylindrical angle-ply specimen. For this layup $A_{16}^{\circ} = A_{26}^{\circ} = 0$. Because two layers are not sufficient for symmetry through the thickness $B_{ij} \neq 0$ and K_1, K_2 are real non-zero quantities. There will then be coupling between the layers of a laminate which contribute to the total response in a manner similar to the role played by the shear coupling terms $\bar{S}_{16}$ and $\bar{S}_{26}$. Therefore, the application of a tensile stress $h\sigma_x = N_x$ will give

$$\begin{aligned} \varepsilon_x^0 &= A_{11}^{\circ} N_x \\ \varepsilon_y^0 &= A_{12}^{\circ} N_x \\ \varepsilon_{xy}^0 &= A_{66}^{\circ} K_1 N_x \end{aligned}$$

Thus, a comparison of this result with the earlier illustration of a tensile traction N_x applied to the unidirectional tube clearly illustrates the complication which can be incurred if the lamination sequence is not properly executed. In addition there does not exist a simple conversion between the tensorial moduli and conventional engineering constants when the coupling term, B_{ij}, is non-zero.

It may seem to the uninitiated that equation (5-59) is a relatively simple discipline of a filament wound cylinder. In fact, once the reader recognizes that filament winding is a fabrication process in which successive windings of filaments or tapes along helical trajectories are made, the source of the complications are recognized. The windings form laminaes which contain, alternately, left and right hand helics of angle (θ); an unsymmetrical angle ply $[-\theta/\theta]_n$ laminate. This laminate was seen in Figures 3-5 and 3-8. For such a laminate the following terms in equations (3-40) are zero

$$B_{11} = B_{22} = B_{12} = B_{66} = 0$$

and

$$D_{16} = D_{26} = A_{16} = A_{26} = 0$$

This means B_{16} and B_{26} are finite and complicate the response. As the number of alternating layers increase, the A and D terms will go the balanced and symmetrical values with the 16, 26 terms going to zero. This effect is demonstrated in Figure 5-6 for a filament wound glass-epoxy system. The "curling under" of the ring is induced by the cure shrinkage in the hoop direction which is equivalent to the ± 30 laminate; see Figure 3-8 and Chapter 6.

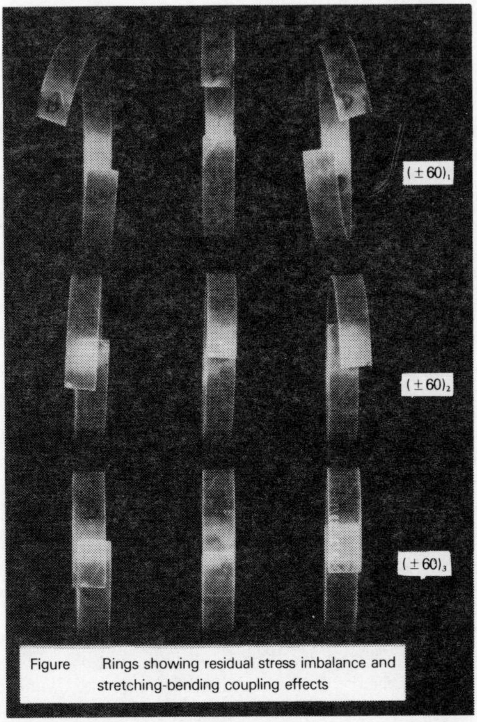

Figure 5-6. *Glass-epoxy rings cut from (± 60) filament wound tubes showing residual stress inbalance and stretching-bending coupling effects (H. Briscall, AWRE, England).*

NOTATION, CHAPTER 5

Notation of Chapter 3 plus

X, θ, Z = Cylindrical coordinate system
U = Airy stress function
r = radius

REFERENCES

General Laminated Plates

1. Smith, C. B., "Some New Types of Orthotropic Plates Laminated of Orthotropic Material," *Journal of Applied Mechanics, 20,* 286–288 (June 1953).
2. Reissner, E. and Stavsky, Y., "Bending and Stretching of Certain Types of Heterogeneous Aeolotropic Elastic Plates," *Journal of Applied Mechanics, 28,* 402–408 (September 1961).
3. Stavsky, Y. and McGarry, F. J., "Investigation of Mechanics of Reinforced Plastics," Air Force Technical Documentary Report No. Wadd-TR-60-746 (June 1962).
4. Dong. S. B., Matthiesen, R. B., Pister, K. S., and Taylor, R. L., "Analysis of Structural Laminates," Air Force Report ARL-76 (September 1961).
5. Whitney, J. M., "Bending-Extensional Coupling in Anisotropic Laminated Plates," *Journal of Composite Materials* (January 1969).
6. Whitney, J. M., "A Study of the Effects of Coupling Between Bending and Stretching on the Mechanical Behavior of Layered Anisotropic Composite Materials," Ph.D. Dissertation, The Ohio State University (August 1968).
7. Ashton, J. E., "Approximate Solution for Unsymmetrically Laminated Plates," *Journal of Composite Materials* (January 1969).

Mid-Plane Symmetric Laminated Plates and General Anisotropic Plates

8. Thielman, W., Contribution to the Problem of the Bulging of Orthotropic Plates, F. G. H. Report No. 150/19.
9. Lekhnitskii, S. G. *Anisotropic Plates.* 2nd Edition, Moscow-Leningrad: OGIZ (1947) English edition translated by S. W. Tsai and T. Cheron, Gordon and Breach (1968).
10. Ashton, J. E. and Waddoups, M. E., "Analysis of Anisotropic Plates," *Journal of Composite Materials* (January 1969).
11. Ashton, J. E. and Whitney, J. M. *Theory of Laminated Plates.* Lancaster, PA: Technomic Pub. Co. (1970).
12. Whitney, J. M., Browning, C. E., and Mair, A. *Composite Materials: Testing and Design.* (Third Conference) ASTM STP 546.
13. Timoshenko, S. *Strength of Materials.* Third edition, New York: D. Van Nostrand, Inc., pp. 170–175 (1955).
14. Whitney, J. M. and Pagano, N. J., *J. of Appl. Mech., 37,* 1031 (1970).
15. Hoff, N. J. in *Engineering Laminates,* A. G. H. Dietz. New York: J. Wiley and Sons (1949).
16. Pagano, N. J., *J. of Comp. Mat., 1,* 336 (1967).

Shells

4. Dong. S. B., Matthiesen, R. B., Pister, K. S., and Taylor, R. L., "Analysis of Structural Laminates," Air Force Report ARL-76 (September 1961).
17. Novozhilov, V. V. *Thin Shell Theory.* 2nd Edition, Groningen, The Netherlands: P. Noordhoff Ltd. (1964).
18. Donnell, L. H., "Stability of Thin-Walled Tubes Under Torsion," N.A.C.A. Report No. 479 (1933).
19. Tasi, J., "Effect of Heterogenity on the Stability of Composite Cylindrical Shells Under Axial Compression," *AIAA Journal, 4*(6) (1966).
20. Whitney, J. M. and Halpin, J. C., "Analysis of Laminated Anisotropic Tubes Under Combined Loading," *Journal of Composite Materials* (July 1968).

6
Structure Property Relationships for Composite Materials

6.1 INTRODUCTION

THE OBJECTIVE OF THIS CHAPTER IS THE DEVELOPMENT OF PROCEDURES for cause and effect relationships between the component constituent properties, their geometrical and volumetric distribution in space, and the properties of a lamina. Accordingly, the analysis necessarily involves the study of a heterogeneous material consisting of reinforcements embedded in a matrix material. The overall response of a ply or thin sheet of the composite containing many parallel filaments in a representative volume element, is thereby defined and the ply is subsequently treated as being homogeneous. As indicated in earlier chapters, composite structures are normally fabricated by a lamination process in which the plies are oriented in a predetermined manner. The laminate properties are then defined in terms of the characteristics of a single ply—Chapters 2 and 3. Since these structures are usually categorized as plates or shells, the medium for accomplishing this objective is the classical laminated plate or shell theory of Chapters 3 and 5. By employing this theory, boundary value problems can be formulated and solved for such diverse topics as stiffness, strength, vibrational and stability characteristics of composite structures. Hence, the results of this chapter, when combined with laminated plate or shell theory constitutes a continuous flow or link between the mechanical properties of the constituents of a composite and the response of structural members of practical importance.

In this chapter an extensive review of the current state of analysis shall be presented. Engineering approximations, justified by the results of more rigorous approaches, will be introduced and then several illustrative example problems will be worked out.

6.2 REVIEW OF REQUIRED MATERIAL COEFFICIENTS

In Chapter two the constitutive equation for an orthotropic lamina was derived and it was shown; eqs. (2-28) and (2-31) that the stiffness properties of a lamina can be characterized by specifying the values of E_{11}, E_{22}, ν_{12}, and G_{12}. In some applications or calculations, the properties of the lamina in the thickness direction are of concern. Fiber reinforced systems fabricated from bundles, or tons, of fibers result in random fiber nesting in the 2-3 plane. Accordingly, such a material appears to be isotropic in the 2-3 plane. Such a material is denoted by the phrase, transversely isotropic, and introduces a new independent elastic constant

$$G_{23} = \frac{E_{33}}{2(1 + \nu_{23})} \tag{6-1}$$

where G_{23} is the transverse shear moduli in the 2-3 plane perpendicular to the fiber direction. The Poisson ratio ν_{23} also characterizes the response in the 2-3 plane as well as the transverse moduli, E_{33}, but these three constants are interrelated through the isotropic relationship (5-1). The corresponding compliance relationships are

$$\begin{aligned}
&S_{11} = \frac{1}{E_{11}} && S_{22} = \frac{1}{E_{22}} && S_{33} = \frac{1}{E_{33}} \\
&S_{44} = 1/G_{23} && S_{12} = -\frac{\nu_{12}}{E_{11}} = -\frac{\nu_{12}}{E_{22}} && \\
&S_{55} = 1/G_{13} && S_{13} = -\frac{\nu_{31}}{E_{33}} = -\frac{\nu_{13}}{E_{11}} && \\
&S_{66} = 1/G_{12} && S_{23} = -\frac{\nu_{23}}{E_{22}} = -\frac{\nu_{32}}{E_{33}} && \\
&S_{55} = S_{66} && S_{22} = S_{33} && S_{12} = S_{13}
\end{aligned} \tag{6-2}$$

Knowledge of these parameters (including knowledge of geometric transforms and lamination sequence) specify both lamina and laminate properties. Accordingly, it is these coefficients which we must predict from a knowledge of the reinforcement and matrix characteristics. The transverse properties ν_{23}, G_{23}, G_{12}, and E_{22} are the most difficult coefficients to predict but are of great value as the solution of the G_{23} problem leads, by analogy, to the prediction of such transport properties as dielectric constants, electrical and heat conduction, diffusion coefficients, etc. These transport coefficients possess the properties of second order tensors—for example the thermal conduction coefficients are expressed as

$$\begin{matrix} K_1 & 0 & 0 \\ 0 & K_2 & 0 \\ 0 & 0 & K_3 \end{matrix} \tag{6-3}$$

For oriented fiber composites, the material is treated as transversely isotropic, or $K_2 = K_3$. Transformation equations for a matrix possessing the structure of (6-3) similar to those developed in Chapter 2 or Appendix A, are expresed as

$$
\begin{aligned}
\bar{K}_1 &= m^2K_1 + n^2K_2 \\
K_2 &= n^2K_1 + m^2K_2 \\
K_{12} &= mn(K_2 - K_1)
\end{aligned} \tag{6-4}
$$

where

$$m = \cos\theta \text{ and } n = \sin\theta$$

In addition to the transport coefficients and elastic moduli it will be necessary for us to characterize the expansion properties of a lamina: e_1, e_2, and e_3. For oriented fibrous composites $e_2 = e_3$ and is described by the formulations indicated in (6-3) and (6-4). A knowledge of e_1 and e_2 is indispensable in estimating the mechanical consequence of temperature variations in service conditions, consequences of molding or curing shrinkage, effects of swelling agent upon composite response, etc. Problems of this type are formulated with the aid of the Duhamel-Neumann form of Hooke's law developed in Chapter 2, equation (2-41):

$$\varepsilon_i = S_{ij}\sigma_j + \alpha_i \Delta T \tag{6-5}$$

$$\sigma_i = C_{ij}(\varepsilon_j - \alpha_j \Delta T) \tag{6-6}$$

In what follows we shall develop relationships between the composite coefficients for oriented continuous and chopped reinforcements in either oriented or random geometries.

6.3 REVIEW OF FORMAL APPROACHES TO THE PREDICTION OF ELASTIC CONSTANTS [6-1]

Several basic assumptions are common to all treatments treated in this section: (1) the ply is macroscopically homogeneous, linearly elastic, and generally orthotropic or transversely orthotropic; (2) the fibers are linearly elastic and homogeneous; (3) the matrix is linearly elastic and homogeneous; (4) fiber and matrix are free of voids; (5) there is complete bonding at the interface of the constituents and there is no transition region between them; (6) the ply is initially in a stress free state; and (7) the fibers are (a) regularly spaced and (b) aligned. While these assumptions are admittedly severe, they do yield results which are useful for technological problems.

A. Self-Consistent Model Methods

A rigorous approach to predict the elastic constants is the "self-consistent method" of which there are two basic variants, namely, the method by Hill [6-2] and that used by Kilchinskii [6-3,4] and Hermans [6-5]. These are discussed below.

Hill [6-2] used the method, proposed by Hershey [6-6] and Kroner [6-7] for aggregates of crystals, to derive expressions for the elastic constants. Making all the basic assumptions, except (7a), he modeled the composite as a single fiber embedded in an unbounded homogeneous medium which is macroscopically indistinguishable from the composite (basic assumption 1). The medium is subjected to a uniform loading at infinity which induces a uniform strain field in the filament (for proof of this statement see [6-8]). This strain field is used to estimate the elastic constants. Hill showed that the expressions derived by this method give reliable values at low filament volume ratios, reasonable values at intermediate ratios, and unreliable values at high ones.

The other variant of the self consistent method, proposed by Frohlich and Sack [6-9] for predicting the viscosity of a Newtonian fluid containing a dispersion of equal elastic spheres, consists of modeling the composite as three concentric cylinders, the outer one being unbounded. The innermost cylinder is assumed to have the elastic properties of the filaments, the middle one those of the matrix, and the outermost has the properties of the composite. The solid is subjected to homogeneous stresses at infinity. The resulting elastic fields are determined and they are used to get the values of the elastic constants. In applying this method, Kilchinskii [6-3,4] made the further assumption that the filaments are distributed in a regular hexagonal array. Approximating the outer radius of the matrix cylinder by requiring that the cross-section areas of the basic hexagon and the matrix cylinder be the same, he obtained expressions for the composite elastic constants and coefficients of thermal expansion.

Hermans [6-5] relaxed basic assumption (7a) and used the condition that the ratio of the volume of the fiber to that of the cylinder containing the matrix and fiber be equal to v_f in the composite to determine the radius of the matrix cylinder. Allowing both the matrix and filaments to be transversely isotropic, he obtained expressions for the elastic constants. For isotropic filaments and matrix, Hermans' results reduce to those given by Kilchinskii because the radius of the matrix cylinder is the same in both cases. Ebert, et. al [6-10] proposed a model in which the role of the matrix and filaments are interchanged at high filament contents. Using this model they obtained expressions for E_1.

Whitney and Riley [6-11] used a model consisting of a single filament embedded in a matrix cylinder of finite outer radius to derive equations for E_{11}, E_{22}, and G_{12}. The analysis consists of finding the stresses in the cylinders for various surface loadings and subsequently, using the results in an energy balance. The results for E_{11} and G_{12} obtained by using this model are in agreement with those of Kilchinskii and Hermans; however, the expression for E_{22} is different. Whitney [6-12] extended this method to transversely isotropic fibers. He also proposed a method for taking into account the effect of the fiber twist on E_{11} [6-13].

B. Variational Methods

The energy theorems of classical elasticity have been used by several authors to obtain bounds on the ply's elastic properties. The minimum complementary energy theorem yields the lower bound, while the minimum potential energy theorem yields the upper bound. The paper pertinent to filamentary composites are reviewed below.

Paul [6-14] derived bounds for E_{11} (or E_{22}), G_{12}, and ν_{12} by treating the composite as being transversely isotropic. He assumed that both of the constituents are in a state of homogeneous stress (or strain) in the macroscopic sense. Paul's bounds are far apart particularly at immediate ν_f values.

Hashin and Rosen [6-15] considered both the hexagonal and random arrays of fibers. They assumed the ply to be transversely isotropic and apparently were the first to introduce the concept of restrained matrix. Using potential and complementary energy theorems, Hashin and Rosen derived both bounds for all five elastic constants. In the random array case, the bounds for G_{12} coincide. Comparison of their random-array results with experimental data shows that there is poor agreement for G_{12} and that the experimental results lie fairly close to the upper bound for E_{22}. The variable $\partial \nu$ should be replaced by $1-\partial \nu$ in equation (28) (see [6-15] and [6-17]). Hashin [6-18] extended this method to obtain bounds for the bulk and the two shear moduli for plies with constituents of arbitrary phase geometry. These bounds can be far apart particularly for composite of the boron-epoxy type. No comparison with experiment is given in [6-17].

Hill [6-19] derived bounds for the ply elastic constants that are analogous to those of Hashin and Rosen [6-15]. He showed by formal and physical arguments that these are the best bounds that can be obtained without taking into account detailed local geometry and that the rule of mixtures is the lower bound for E_{11}. In his formulation he relaxed basic assumption (7) and considered a mathematical model of a large number of filaments so that any local strain or stress irregularities have negligible effects on the average stress or strain over the cross-section of the model. No numerical results are given.

Levin [5-10] used Hill's approach to obtain bounds on the ply coefficients of thermal expansion. It is noted that equation (3-12) of [6-20] is incorrect. Schapery [6-21] independently obtained similar bounds for a multiphase system.

C. Exact Methods

An exact method of analyzing filamentary composites consists of assuming that the fibers are arranged in a regular periodic array. In this method, the resulting elasticity problem is solved either by a series development or by some numerical scheme. The resulting elastic fields are averaged to get expressions for the desired elastic constants.

This method appears to have been first used by Fil'shtinskii et. al [6-22,23] to predict the reduced elastic moduli of a plate weakened by a doubly periodic array of equal circular holes. Using the classical Kolosov-Muskhelishvili stress functions, and the periodicity properties of the Weierstrauss zeta and special meromorphic functions, they reduced the problem to the solution of an infinite set of algebraic equations. They proved the convergence of this solution and obtained exact and approximate expressions for the elastic constants. In later papers, they extended the analysis to a plate containing rigid [6-24] and elastic circular inclusions [6-25]. Even though the analysis was performed for plane stress, the results are applicable to the case of plane strain, and hence to fibrous composites.

Van Fo Fy [6-26] applied Fil'shtinskii's [6-26] formulation to filamentary composites. Assuming that the fibers are isotropic and that they form a regular doubly-periodic array, he found approximate expressions for the ply elastic constants. In [6-27], he presented a detailed solution of the problem for shear loading giving exact and approximate expressions for G_{12}, G_{23}, and G_{31}. Exact expressions for the other moduli are given in [6-28]. Graphical results with detailed discussion are presented in [6-27,28]. Exact and approximate expressions for the coefficients of thermal expansion are derived in [6-30]. Composites containing hollow fibers are considered in [6-31]. Exact expressions for the thermal conductivities of plies containing hollow and solid fibers are derived in [6-32]. An attempt to take into account the interface layer between fibers and matrix is made in [6-33]. Limited comparison is made with experimental data. Van Fo Fy claims that the approximate expressions are in good agreement (less than ten percent difference) with the exact ones. As a general comment, it may be said that the papers are hard to follow if taken individually because the notation is not always defined in a given paper.

An alternate approach consists of using symmetry arguments to reduce the doubly-periodic elasticity problem to one for a finite region, called the fundamental repeating element. The resulting problem is then solved by various techniques.

The earliest application of this approach appears to be due to Hermann and Pister [6-34] who obtained the elastic and thermoelastic constants of the ply for a square array of fibers. Taking full advantage of symmetry, they reduced the problem to one for a square containing a circular inclusion. Solving the reduced problem, they obtained results for the five elastic constants, two coefficients of linear thermal expansion, and two heat conductivities. They also investigated the sensitivity of the ply's thermo-elastic properties to variation of the properties of the constituents. No details of the solution to the reduced problem are given.

Wilson and Hill [6-35] used a rectangular fundamental element with a hole (rigid inclusion) in the center to solve the problem for plate containing a rectangular array of doubly-symmetric holes (rigid inclusions). The resulting problem was also solved by using complex variables methods and mapping the elements into two concentric cylinders. In their analysis, the boundary conditions at the hole (rigid inclusion) are satisfied exactly, while those at the rectangular sides are satisfied approximately by a point matching procedure. Results were also obtained for elastic fibers in an elastic matrix [6-36].

Adams and Doner [6-37,38] solved the problem for the rectangular fundamental region by a finite-difference scheme. They give extensive results for G_{12} [6-37] and E_{22} [6-38]. Results for the other elastic constants and the coefficients of thermal expansion are given in [6-36].

Pickett [6-39,40] and Chen and Cheng [6-41] considered a hexagonal array of fibers and took a triangular fundamental element. Using a series solution in polar coordinates, they satisfied the boundary condition at both the fiber-matrix interface and the sides of the triangle by a modified least square method. They obtained results for all the elastic constants. Pickett [6-40] also used the rectangular element to get results for a tape containing a single row and a doubly-periodic rectangular array of fibers.

Clausen and Leissa [6-42] used a triangular element in their consideration of the doubly-periodic array of symmetric fibers. Using a series solution to solve the resulting problem, they satisfied the boundary conditions at both the fiber-matrix interface and symmetry lines by a point matching procedure in the least square sense.

Bloom and Wilson [6-43] used the best features of [6-35,42] in their consideration of the hexagonal array. In this analysis, they used a triangular fundamental element and solved the problem by an infinite series. The series was selected in such a manner that the boundary conditions at the fiber-matrix interface are satisfied identically. The conditions at the symmetry lines are satisfied approximately by a point matching technique. They obtained results for E_{11} and G_{12} (the analysis was also reported in [6-36]). Quackenbush and Thomas [6-44] extended this analysis to transversely isotropic fibers.

Foye [6-45,46] was the first to apply the discrete element method for the prediction of E_{22}, G_{12}, ν_{12}, and ν_{23}, by employing a triangular constant strain discrete element. He investigated square, diamond, and hexagonal arrays and fibers of various cross sectional areas.

An extensive treatment of periodic arrays is found in Grigolyuk and Fil'shtinskii [6-47]. See also, Lomakin and Koltumov [6-48], Haener [6-49], Piehler [6-50], and Meijers [6-51] for closely related work.

Behrens [6-52,53] investigated the propagation of plane sound waves in fibrous composites to determine the ply elastic constants and the ply heat conductivities. The central point in this method is that the equations of motion of the sound waves are solved for long wave lengths and the results are subsequently compared with corresponding solutions for sound waves in orthorhombic crystals. The governing assumption is that the wave lengths are very long as compared with interspacial fiber distances.

Springer and Tsai [6-54] derived equations for the ply heat conductivities based on rectangular and square array models. The equations for K_1 are derived from the rule of mixtures; while those for K_2 are derived by establishing the analogy between shear stress and temperature field at a local point.

Shlenskii [6-54] derived an expression for K_2 taking into account the void content. He investigated the validity of his simplifying assumptions by comparing the derived values of K_2 with those given by an electrical analogy. The values of K_2 given by the electrical analogy and the derived formula are in good agreement with his experimental results. He emphasizes the importance of void content. More recently, Vishnevskii and Shlenskii [6-55] derived expressions for K_1 and K_2 for a ply with curved filaments. They assumed that the filament curvature varies sinusoidally along the longitudinal axis.

6.4 THE HALPIN-TSAI EQUATIONS

The product of the above cited efforts is contained in a series of curves (Figures 6-1,2) computed as a function of material parameters. While these curves are of great value they are limited in that only a few cases are available in graphical form. For design purposes it is often desirable to have simple and rapid computational procedures for estimating the ply properties. A desirable situation is one in which simple but approximately precise formula would be available to interpolate the existing exact machine calculations available in the current literature.

Such interpolation procedures have been developed by Halpin and Tsai [6-56] who were able to show that Hermans' solution [6-5] generalizing Hill's self consistent model [6-2] can be reduced to the approximate form:

$$E_{11} \cong E_f v_f + E_m v_m \tag{6-7}$$

and

$$\nu_{12} \cong \nu_f v_f + \nu_m v_m \tag{6-8}$$

and

$$\frac{\bar{p}}{p_m} = \frac{(1 + \zeta \eta v_f)}{(1 - \eta v_f)} \tag{6-9}$$

where

$$\eta = (p_f/p_m - 1)/(p_f/p_m + \zeta) \tag{6-10}$$

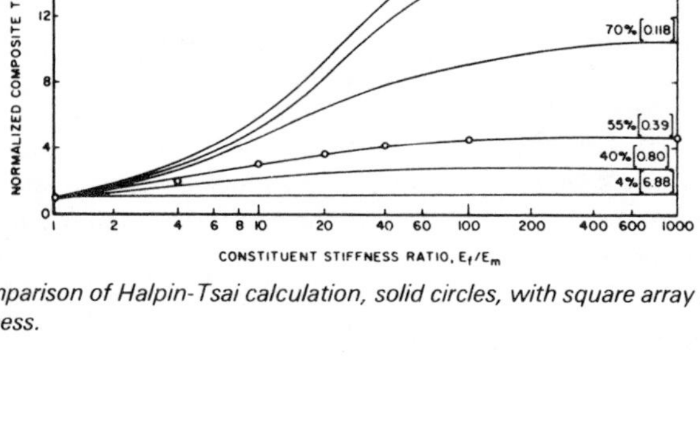

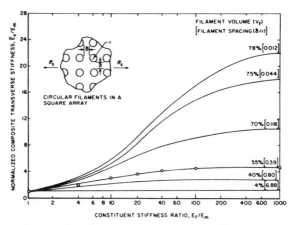

Figure 6-1. *Comparison of Halpin-Tsai calculation, solid circles, with square array calculations for transverse stiffness.*

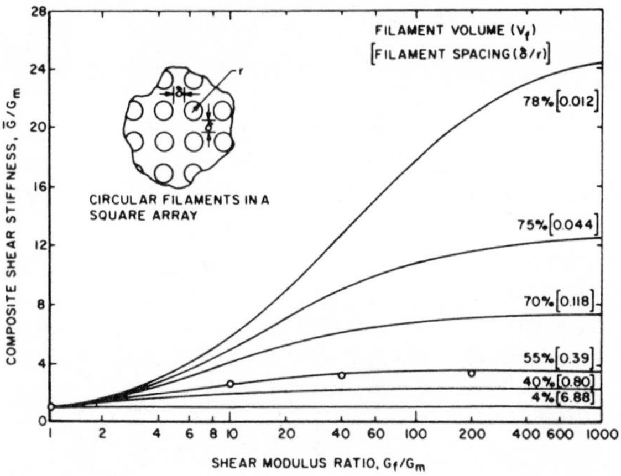

Figure 6-2. *Comparison of Halpin-Tsai calculation, solid circles, with Adams and Doner's square array calculations for longitudinal shear stiffness.*

In this formulation the quantities $\bar{p}$, p_f, p_m, and ζ are identified as:

$\bar{p}$ = composite moduli, E_{22}, G_{12}, or G_{23};
p_f = corresponding fiber modulus, E_f, G_f, ν_f respectively;
p_m = corresponding matrix modulus, E_m, G_m, ν_m;
ζ = a measure of reinforcement which depends on the boundary conditions.

Once the ζ factors are known for the geometry of inclusions, packing geometry and loading conditions, the composite elastic moduli for fiber, tape, and particulate composites are approximated from the Halpin-Tsai formulas. Reliable estimates for the ζ factor can be obtained by comparison of equations (6-9) and (6-10) with the numerical micromechanics solutions employing formal elasticity theory. For example, Figures 6-1 and 6-2 show the predictions of equations (6-9) and (6-10) for various reinforcements/matrix stiffnesses and is compared with the results of Adams, and Doner [6-37,38]. The approximate formula does duplicate their results for all ratios of p_f/p_m. In Figures 6-3 and 6-4 we have a comparison of the dependence of composite moduli, based upon Foye's calculations [6-45,46], as a function of volume fraction for square fibers in a diamond array. For the predictions of shear G_{12}, ζ=1 and for stiffness E_{22}, ζ=2 for the calculations shown in Figures 6-1–6-4. In Figures 6-5 and 6-6 we show the results of our generalized formula compared against the results of different micromechanics calculations. Note that our approximate formula provides a good correlation with Foye's computations for hexagonal array of circular fibers up to ν_f = 65 percent. At higher volume fractions Hewitt and deMalherbe [6-57] have shown that the following modification adjusts for this effect:

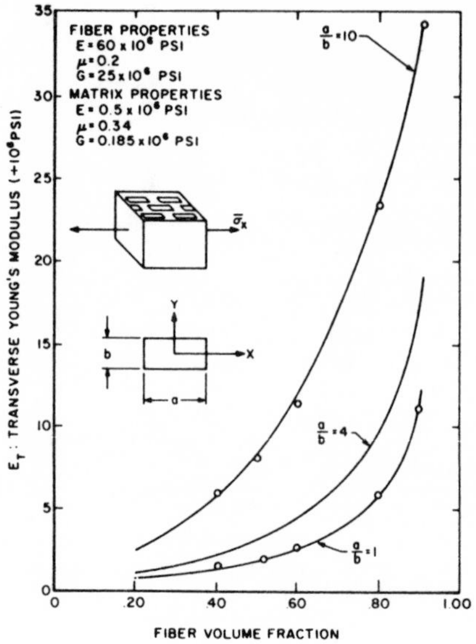

Figure 6-3. *Comparison of Halpin-Tsai calculation, solid circle, with Foye's calculations for the transverse stiffness of a composite containing rectangularly-shaped fibers.*

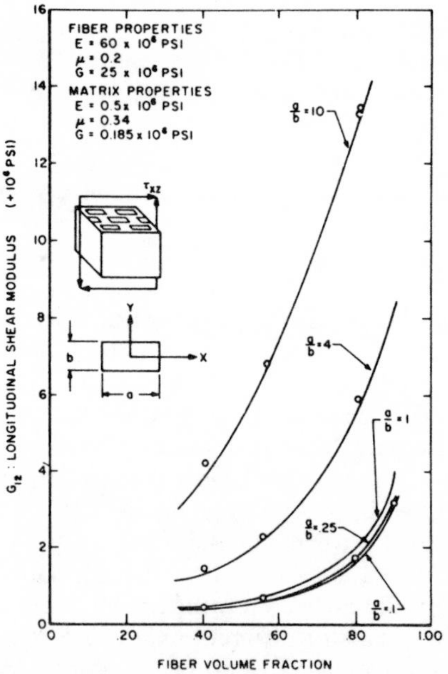

Figure 6-4. *Comparison of Halpin-Tsai calculation, solid circle, with Foye's calculation for longitudinal shear stiffness of composite containing rectangularly-shaped fibers.*

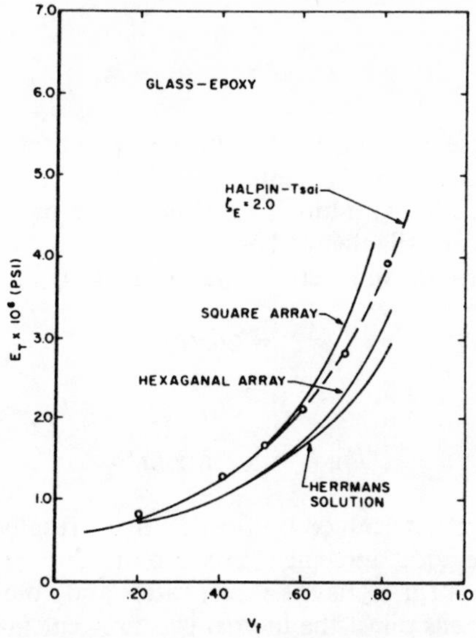

Figure 6-5. *Comparison of various calculations for the transverse stiffness of a glass-epoxy composite.*

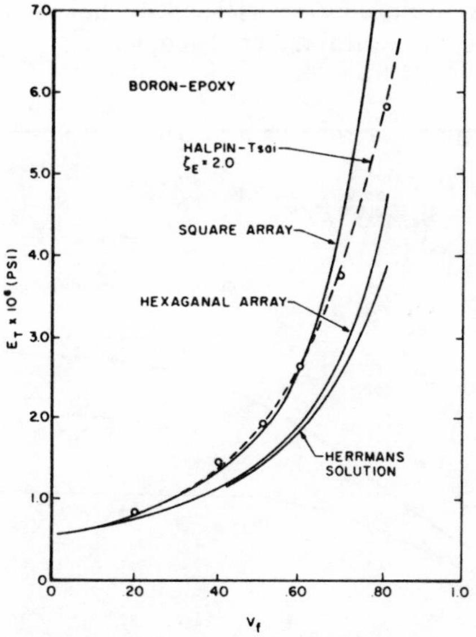

Figure 6-6. *Comparison of various calculations for the transverse stiffness of a boron-epoxy composite.*

$$G_{12} = 1 + 40\, v_f^{10} \tag{6-11}$$
$$E_{22} = 2 + 40\, v_f^{10} \tag{6-12}$$

Further confidence can be gained in this interpolation procedure by returning to Figures 6-3 and 6-4 and examine the dependence of the transverse stiffness moduli and the longitudinal-transverse shear moduli upon increasing aspect ratio for square filaments. The factors ζ_E and ζ_G are functions of the width to thickness ratios and were found [6-58] to be of the form:

$$\zeta_E = 2(a/b) \tag{6-13}$$

and

$$\log \zeta_G = \sqrt{3} \log (a/b)$$

These ζ factors were determined by fitting Foye's results. Results in Figures 6-3 and 6-4 indicate good agreement between our approximate formulas and elasticity calculations for different aspect ratios and volume fractions. These results allow the extension of the limited elasticity calculations to the prediction of, for example, the dependence of the moduli, E_{22} and G_{12}, on the aspect ratio, reinforcement stiffness/matrix stiffness and volume fraction loading as illustrated in Figures 6-7 and 6-8.

The ζ term is a parameter which reflects reinforcement geometry and modulus influences on the elastic coefficients. These ζ and η terms are summarized below for some typical material geometrics:

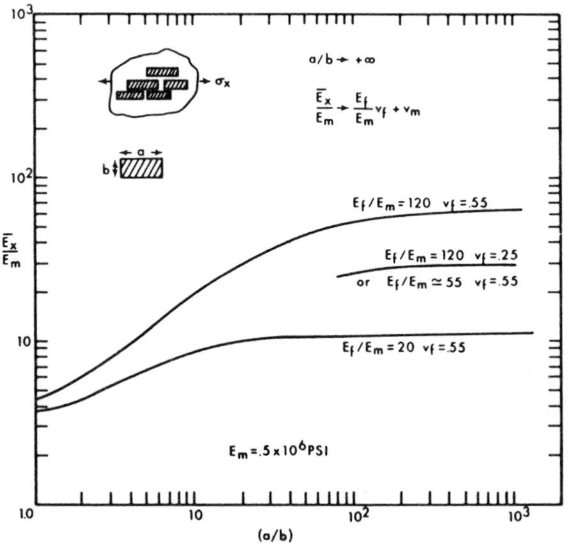

Figure 6-7. Composite transverse stiffness versus aspect ratio for ribbon-shaped reinforcements.

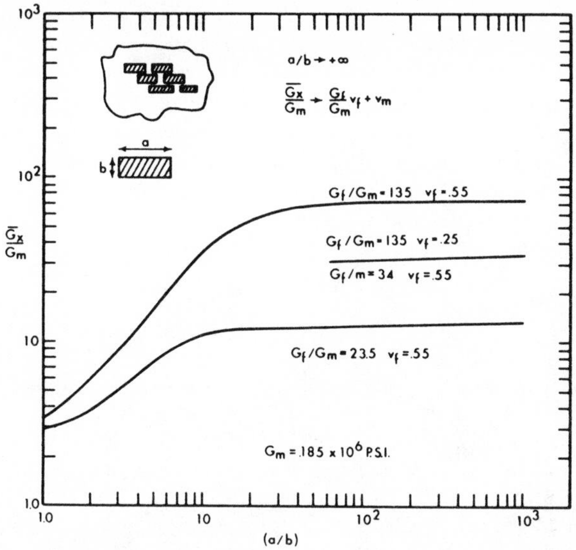

Figure 6-8. *Composite longitudinal shear stiffness versus aspect ratio for ribbon-shaped reinforcements.*

(a) Particulate Composites:

$$\zeta_G \cong \zeta_{G_{12}} \qquad \text{equation (6-11)}$$

$$\zeta_E \cong \zeta_{E_{22}} \qquad \text{equation (6-12)}$$

$$G/G_m \text{ and } E/E_m \qquad \text{equation (6-9)}$$

(b) Voids, a foam, a porous solid:

$$p_f/p_m \to 0$$

therefore

$$\eta_G = 1/\zeta_G;\ \eta_E = 1/\zeta_E$$

$$G/G_m \text{ and } E/E_m \qquad \text{equation (6-9)}$$

(c) Oriented continuous fibers:

$$\zeta_{G_{12}} \text{ and } \zeta_{E_{22}} \qquad \text{equations (6-11) and (6-12)}$$

$$\zeta_{G_{23}} \cong \frac{1}{4-3\nu_m} \tag{6-14}$$

$$\nu_{23} \cong \frac{(1-\eta_{G}v_f)}{(1-\eta_{E}v_f)} \frac{(1+\eta_E\zeta_E v_f)}{(1+\eta_G\zeta_E v_f)} \frac{E_m}{2G_m} - 1 \qquad (6\text{-}15)$$

where ζ_G and ζ_E are equations (6-11) and (6-12) and η_G and η_E is equation (6-10) using ζ_G and ζ_E respectively.

E_{11} equation (6-7)

ν_{12} equation (6-8)

G_{12}/G_m, G_{23}/G_m, and E_{22}/E_m equation (6-9)

(d) Oriented discontinuous fibers:

ℓ/d = longitudinal aspect ratio; length to diameter or thickness

$\zeta_{E_{11}} = 2(\ell/d) + 40\, v_f^{10}$, equation (6-13)

$\zeta_{G_{12}}$ equation (6-11)

$\zeta_{G_{23}} \sim$ equation (6-14)

$\zeta_{E_{22}}$ equation (6-12)

$\nu_{12} \sim$ equation (6-8)

$\nu_{23} \sim$ equation (6-15)

E_{11}/E_m, E_{22}/E_m, G_{12}/G_m, and G_{23}/G_m equation (6-9)

(e) Oriented Continuous ribbon or lamellar-shaped reinforcement:

ω/t = transverse aspect ratios; width to thickness

$\zeta_{E_{22}} = 2(w/t) + 40\, v_f^{10}$, equation (6-13)

$\zeta_{G_{12}} = (\omega/t)^{1.73} + 40\, v_f^{10}$, equation (6-13)

$\zeta_{G_{23}} \sim$ equation (6-12)

$\zeta_{E_{33}} \sim$ equation (6-12)

$$\nu_{23} \cong \left(\frac{v_f}{\nu_f} + \frac{(1-v_f)}{\nu_m}\right)^{-1} \qquad (6\text{-}16)$$

E_{11} equation (6-7)

$$\nu_{12} \qquad \text{equation (6-8)}$$

$$G_{12}/G_m,\ G_{23}/G_m,\ E_{22}/E_m,\ E_{33}/E_m \qquad \text{equation (6-9)}$$

(f) Oriented discontinuous ribbon or lamella-shaped reinforcement; a flake:

$$\zeta_{E_{11}} = 2(\ell/t) + 40\ \upsilon_f^{10}, \qquad \text{equation (6-13)}$$

$$\zeta_{E_{22}} = 2(\omega/t) + 40\ \upsilon_f^{10}, \qquad \text{equation (6-13)}$$

$$G_{12} = ((\ell+\omega)/2t)^{1.73} + 40\ \upsilon_f^{10} \qquad \text{equation (6-13)}$$

$$E_{11}/E_m,\ E_{22}/E_m, \text{ and } G_{12}/G_m \qquad \text{equation (6-9)}$$

remaining elastic constants as in (e) above.
The above cited equations have been verified against experimental data in References [6-58 thru 6-64].

The above cited expressions when combined with the laminated plate theory provide a simple approach for material engineering estimates. In general, these estimates will vary from experimental observations by about 5 to 7 percent.

In closing this section it is constructive to note that there is a large class of materials in which the reinforcements are randomly distributed in the plane of a sheet or plate. Such a sheet of material will conform to plane stress treatments and characterized as plannar isotropic materials or in laminate terms, as a quasi-isotropic laminate. For example, consider the case when $A_{11} = A_{22}$ and $(A_{22}-A_{22}-2A_{66}) = 0$ and recall that the A_{ij} terms have the same angular dependence as the Q_{ij} terms, equation (2-36). Note that only those terms in equation (2-36) which are independent of functions of angles can describe an isotropic condition:

$$\begin{aligned} \bar{A}_{11} &= \bar{A}_{22} = A_{11} = U_1 \\ \bar{A}_{12} &= \bar{A}_{21} = A_{12} = U_4 \\ \bar{A}_{66} &= \bar{A}_{66} = U_5 \\ \bar{A}_{16} &= \bar{A}_{26} = 0 \end{aligned}$$

Inserting these expressions into equations (3-47 thru 3-50) and noting that $U_5 = \frac{1}{2}\,[U_1 - U_4]$ from equation (2-37) led to the result [6-59]:

$$E = \frac{4U_5(U_1-U_5)}{U_1}$$

$$= \frac{U_1-2U_5}{U_1} \qquad (6\text{-}17)$$

$$G = U_5$$

It can be shown that any balanced, symmetric laminate constructed of $2N$ plies oriented at $\theta = \pi/N$ radians, where $N \geqslant 3$, yields an in-plane stiffness matrix which is isotropic. The term quasi-isotropic is utilized because the bending stiffness matrix, D_{ij}, will not, in general, be isotropic. Both the strength and fracture toughness, Table 4-1, will reach a stable value independent of an increased number of lamina orientations, π/N when $N \geqslant 3$, and laminate orientation θ. In Table 4-1 $[0/\pm 60]_s$ is a $\pi/3$ laminate, $[0/\pm 45/90]_s$ is a $\pi/4$ laminate, etc.

An examination of equation (2-43) for the expansional strains will indicate that the expansional strains are independent of θ whenever $e_x = e_y$. The expansional strain field is isotropic in the lamination plane any time a balanced and symmetric laminate has equal properties $\bar{A}_{11} = \bar{A}_{22}$. This statement is valid even for a laminate which may not be quasi-isotropic with respect to the in-plane stiffness: for example $[0/90]_s$ or $[\pm 45]_s$ laminates. The term quasi-isotropic expansion is used in recognition that the out of plane strain e_3 will generally be different, greater than, the in-plane strain; $e = e_x = e_y$. The quasi-isotropic expansional strain is developed in [6-65] and Section 6.6 as

$$e^0 = W_1 + \frac{2(E_{11} - E_{22})W_2}{E_{11} + (1+2\nu_{12})\,E_{22}} \tag{6-18}$$

where the terms W_1 and W_2 are defined in equation (3-72).

Problem 6-1

What is the moduli of a composite material fabricated from a matrix of $E_m = 0.3 \times 10^6$psi and 40 volume percent of Silicon Carbide platelets, $E_R = 70 \times 10^6$ psi, possessing a ratio of length to thickness of 16/1?

For a crude estimate of the modulus of a flake or platelet reinforced material we may employ equations (6-9)–(6-13). For a material of this type $E_{11} = E_{22}$ and is equal to E_{22} of the equivalent ribbon reinforced composite possessing the equivalent aspect ratio (a/b):

From equation (6-13)

$$\xi_E = 2(a/b) + 40\upsilon_f^{10} = 2(16) + 40(0.4)^{10} = 32$$

using equation (6-10),

$$\eta = \frac{(70/13) - 1}{(70/.3) + 32} = .0877$$

Therefore equation (6-10) yields,

$$\frac{\bar{E}}{E_m} = \frac{1+(32)\,(.877)\,(.40)}{1-(.877)\,(.40)}$$

Table 6.1. Transverse transport properties.

Physical Subject	ϕ	$\mathbf{H} = -\nabla\phi$	k	q
Electrostatics	electric potential	electric field intensity	dielectric constant	electric induction
Magnetostatics	magnetic potential	magnetic field intensity	magnetic permeability	magnetic induction
Electrical Conduction	electrical potential	electric field intensity	electrical conduction	current density
Thermal Conduction	temperature	temperature gradient	heat conduction	heat flux
Diffusion	concentration	concentration gradient	diffusion coefficient	concentration flux

$$g = k\,H$$
$$J_x = -D\,\partial c/\partial x$$
$$F_x = K\,\partial T/\partial x$$
$$\tau_{zx} = G\,\partial w/\partial x$$

$$= 18.8$$

$$E = 5.65 \times 10^6 \ psi$$

This result compares favorably with experiment as indicated in Figure 6-9. A similar procedure is followed if one desires to compute the shear modulus G_{12} or the moduli of continuous fiber reinforcement.

Problem 6-2

There is much discussion today regarding the use of two or more different reinforcements in a composite structure. How would we estimate expected moduli of such a material?

For a general hybrid composite one would generally fabricate the composite as a laminated material possessing different material properties in different material layers. Accordingly one may employ lamination theory as developed in Chapters 3 and 4; which allows one to sum through the thickness in proportion to the fraction of the thickness of materials *a, b, c*,. . . etc., to obtain $(A_{ij})_H$ for the hybrid composite [6-57]:

$$[A_{ij}]_H = \frac{a}{h}[A_{ij}]_a + \frac{b}{h}[A_{ij}]_b +$$

If, however, two or more reinforcements are uniformly mixed in a common matrix one would compute $(p/p_m)_a$ for reinforcement a, then incorporate reinforcement b into the composite employing $\bar{p}_a$ as the "effective matrix moduli."

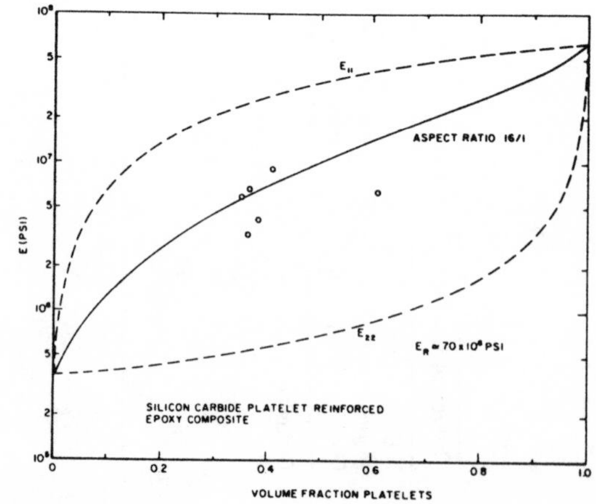

Figure 6-9. Composite Young's modulus for platelet reinforcements.

Problem 6-3

How would you compute the modulus of a random short fiber composite? The analysis would proceed in the following fashion:

1. A random fiber orientation means that the moduli will be isotropic in the plane of a sheet. This situation is described in laminated plate theory as "quasi-isotropic."

$$E \equiv \frac{4U_5(U_1-U_5)}{U_1}$$

$$\nu \equiv \frac{U_1-2U_5}{U_1}$$

$$G \equiv U_5$$

The engineering constants E_{11}, E_{22}, etc. are those representative of the *unidirectionally oriented material.*

2. For a short fiber composite one must now compute the engineering constants. For example, consider a short boron fiber of length, 0.125 in. and a diameter, .004 in., used as reinforcement, 40 percent in an epoxy matrix:

a) Compute E_{11}

$$a/b = \ell/D = \frac{0.125}{.004} = 31$$

$$\xi = 2(32) + 40(0.4)^{10} = 62 \qquad E_f/E_m = \frac{60\times 10^6}{.5\times 10^6} \doteq 120$$

$$\eta = \frac{120-1}{120+62} = 0.655$$

$$\frac{E_{11}}{E_m} = \frac{1+62(.655)(.40)}{1-(.655)(.40)}$$

$$= 23.3$$

$$E_{11} = 11.7\times 10^6 \text{ psi}$$

b) Compute E_{22}

$$\xi = 2(1) + 40(0.4)^{10} = 2$$

Transverse response to short fiber is assumed to be similar to transverse response of continuous fiber.

Following above procedure, we obtain

$$E_{22} = 1.3\times 10^6 \text{ psi}$$

c) Compute E

$$\bar{E} = [3/8(11.7) + 5/8(1.3)] \times 10^6$$
$$= 5.2 \times 10^6 \ psi$$

This crude stiffness estimate can be compared against experiment in Figure 6-10. Mixtures of two chopped fibers can be evaluated by combining problems 6-2 and 6-3. The same procedure would also be applied if one wished to compute the bending stiffness of a beam containing chopped fiber in the mid section and continuous oriented fiber layers on the surface faces.

6.5 TRANSPORT PROPERTIES

Composite materials are frequently employed in engineering applications in which one is concerned with the transport of electrical and/or magnetic fields, electrical conduction, heat conduction, or the permeation of a material by a contained fluid or vapor. Accordingly, we are concerned with the relationship between composite structure and such physical constants as dielectric constants, heat conduction, magnetic permeability, electrical conduction, and the diffusion coefficients. As indicated in section 6.2 these material parameters are second order tensors described by equation in the form of (6-1) and (6-2).

For a fiber reinforced material, the axial coefficient, k_1 is simply expressed as [6-66]

$$k_1 = v_f k_f + v_m k_m \tag{6-19}$$

The transverse coefficient, k_2 can be computed by envoking an analogy from [6-54] classical physics between the in-plane shear field equations and boundary conditions to the transverse transport coefficient, k_2, as indicated in

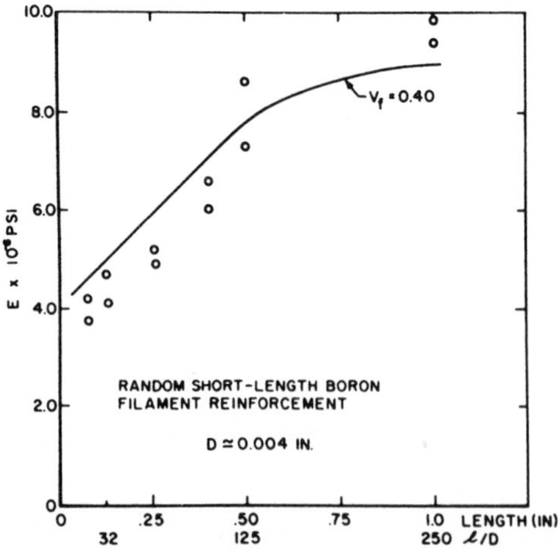

Figure 6-10. Prediction of a random short boron fiber mat reinforced composite.

Table 6.1. Thus, we may compute k_2 for the specific problem of concern by employing equations (6-9), (6-10), and (6-14).

$$\frac{k_2}{k_m} = (1 + \xi\eta v_f)/(1 - \eta v_f) \tag{6-20}$$

$$\eta = (k_f/k_m - 1)/(k_f/k_m + \xi)$$

$$\xi = \frac{1}{4-3\nu_m} = \xi G_{23}$$

Where k_2, k_m, and k_f are the appropriate transport coefficients for composites in the transverse direction, matrix, and fiber.

Problem 6-4

What will be the thermal conductivity perpendicular to plate of a 50 volume percent graphite reinforced plastic matrix composite?

The thermal conductivities of fiber and matrix are

$$k_m = 1.5\ Btu/(hr\text{-}ft^2\text{-}°F/in)$$
$$k_f = 666\ Btu/(hr\text{-}ft^2\text{-}°F/in)$$

For a circular or square fiber cross-section we determine from (6-20)

$$\xi_K = 0.38$$

Accordingly, $k_f/k_m = 666$

$$\eta = \frac{666 - 1}{666 + 0.38} = 0.998$$

$$\frac{k_2}{k_m} = \frac{1 + (0.34)(.998)(.5)}{1 - (.998)(.5)}$$

$$\sim 2.4$$

This estimate is low with respect to the data in Figure 6-11 but is compatible with the results of Gogol and Furmanski [6-67]:

v_f	*Experimental*	*Halpin-Tsai for k_2/k_m*
0.5	2.7	2.4
0.28	1.6	1.52

for systems with $k_f/k_m > 10^3$. For systems where $k_f/k_m \cong 0.1$ the comparison is

υ_f	Experimental	Halpin-Tsai for k_2/k_m
0.5	.35	.32
0.28	.55	.52

Problem 6-5

What is the relationship of diffusion coefficients D_1 and D_2 of a composite to the matrix properties? Assume a fiber in which the fluid, say moisture, is not soluble and has therefore lost a zero diffusion coefficient, $k_f = 0$.

From equation (6-19)

$$D_1/D_m \cong (1-\upsilon_f)$$

From equation (6-20)

$$k_f/k_m = 0$$

$$\eta = -1/\xi_{23}$$

$$\frac{D_2}{D_m} \cong \frac{1-\upsilon_f}{1+\upsilon_f/\xi_{23}}$$

This expression suggests that D_2 will be about ten percent of D_m. This expectation is compatible with experimental observation [6-68, 6-69].

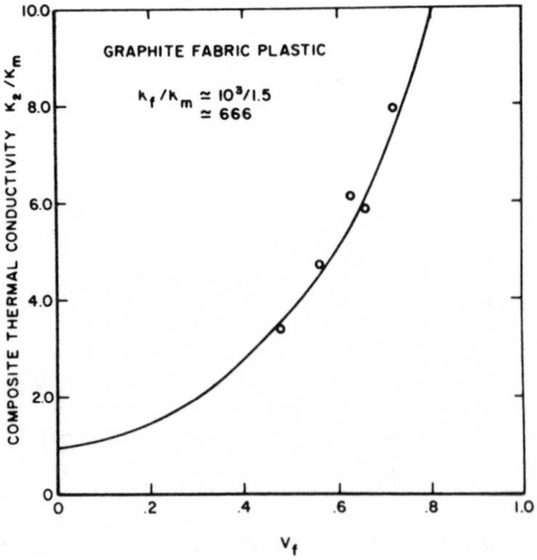

Figure 6-11. Prediction of the transverse thermal conductivity of a fiber reinforced composite.

Problem 6-6

It is desired to fabricate commercial piping or storage tanks with a minimum amount of seepage of the fluid to be contained. What guidance can you provide someone with these procedures?

Of course, one could compute the diffusion of composites from commercially available metal if they are known for the specific materials and fluids of concern. However, one is instinctively aware that few liquids or vapors diffuse readily through glass. Furthermore, we know from equations (6-9)–(6-13) and Figure 6-4 that high-aspect ratio rectangular cross sectional reinforcements produce enormous changes in the inplane shear stiffness. Accordingly equations (6-20) would suggest high aspect ratio (large a/b ratios) glass ribbon-shaped reinforcement [6-57] in piping or storage tank application.

6.6 THERMAL EXPANSION AND DIMENSIONAL STABILITY

The prediction of thermal expansion of unidirectional composites as functions of constituent material properties and phase geometry has received considerable attention in recent years. Schapery [6-21] has derived upper and lower bounds as well as convenient approximate expressions for thermal expansion coefficients. Halpin and Pagano [6-65] have shown that these computations are also valid in any form of dilation strain such as shrinkage during cure, swelling of a composite by an adsorbed liquid or vapor, etc. A unified treatment of these diverse phenomena which effect dimensional stability of fabricated composite objects has been outlined in sections 2.5 and 3.9. The principle expression relating expansional strains to mechanical stresses are equations (2-41)–(2-43), (3-66)–(3-69).

$$\sigma_i = C_{ij}(\varepsilon_j - e_j) \qquad (6\text{-}6)'$$

Employing Schapery's results we may now estimate the longitudinal expansional strain for continuous oriented fibers as

$$e_1 = \frac{E_m e_m v_m + E_f e_f v_f}{E_m v_m + E_f v_f} \qquad (6\text{-}21)$$

and the transverse expansional strain

$$e_2 = e_3 = (1 + \nu_n)e_n v_n + (1 + \nu_f)e_f v_f - e_1(\nu_f v_f + \nu_m v_m) \qquad (6\text{-}22)$$

The dependence of the thermal expansion coefficients or the equivalent expansional straining are illustrated in Figure 6-12. The initial increase in transverse strain or expansion is due to the axial restraint of the fibers. In effect, low expansions are obtained in the axial direction because the fiber, in effect, compresses the matrix and "squeezes" it out in the transverse directions.

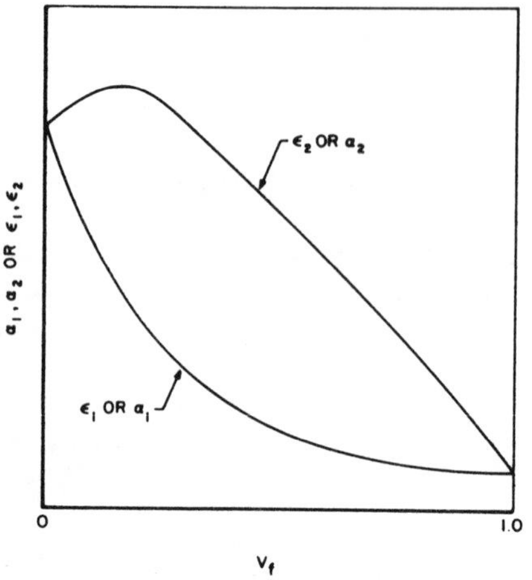

Figure 6-12. Dependence of thermal expansion coefficients and/or expansional strains on volume fraction of reinforcement.

Equations (6-4), (6-21), and (6-22) were experimentally confirmed by Halpin and Pagano [6-59] on a unidirectional material as indicated in Figure 2-14. The reader may check the calculation by employing the following constants:

From	Compute
$E_f = 292{,}000$ PSI	$E_{11} = 132{,}000$ PSI
$E_m = 300$ PSI	$E_{22} = 1{,}050$ PSI
$\nu_f = 0.2$	$\nu_{12} = 0.36$
$\nu_m = 0.4999$	$G_{12} = 263$
$v_f \cong 0.45$	$e_1 \sim 0.01 - .02$
$e_f \cong 0.84$	$e_2 \sim 0.70 - .75$
$e_m \cong 0.01$	

Halpin and Pagano actually employed non-linear strains in their calculations, however, linear calculations as outlined here give reasonable estimates of ply properties.

Once the expansional coefficients (e or α) of a single layer of fiber reinforced material have been computed from constitutive material properties, the response of a laminate can be determined through the use of the laminated plate theory of Chapters 3 and 4 as modified by equation (6-6)′. Assuming a proper stacking or lamination sequence such that no coupling exists through the thickness of a laminate, Halpin and Pagano obtained an exact solution for the expansional properties of a laminate:

$$e_1^0 = \frac{A_{22}R_1 - A_{22}R_2}{A_{11}A_{22} - A_{12}^2} \tag{6-23}$$

$$e_2^0 = \frac{A_{11}R_2 - A_{12}R_1}{A_{11}A_{22} - A_{12}^2} \tag{6-24}$$

where the terms are defined in equations (3-72).

Using the material properties cited above, the laminate expansional strains for angle ply composites were computed and is shown in Figure 2-14. This calculation is somewhat surprising in that it suggests that at certain angles, the laminate will contract in the longitudinal direction when heated or swollen. This prediction was confirmed by Halpin and Pagano as indicated by the triangular points. Note that a longitudinal contraction is compensated by an increased transverse expansion: the expansional strain or coefficient at 90°—Figure 2-14. Calculation for the thermal expansion coefficients of structural composites are shown in Figure 6-13. Calculations of this type have obvious technological significance as they specify the material parameters and geometric construction necessary to yield structural elements with zero or minimal expansional coefficients.

The expansional strain in the thickness direction has been discussed in [6-72] and [6-73]. For balanced and symmetric angle-ply laminates of equal thickness lamina the thickness expansional strain, e_z is given as

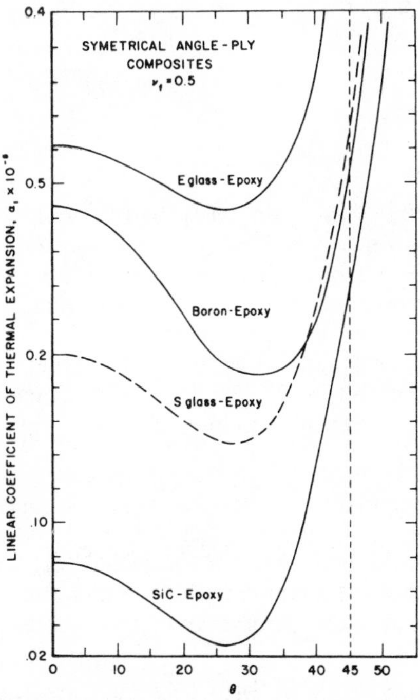

Figure 6-13. *Linear thermal expansion coefficients for typical angle ply composites.*

$$e_z^0 = e_2^0 + \frac{E_{11}\ G_{12}\ (e_2 - e_1)\ (\nu_{23} - \nu_{21})}{E_{11}\ E_{22}\ Cot^2\theta\ +\ G_{12}\ [E_{11}\ +\ (1 + 2\ \nu_{12})E_{22}]} \quad (6\text{-}25)$$

where the material parameters refer to lamina properties. Note that e_z is dominated by the Poisson's effects responding to the in plane strains in the lamina. The $Cot^2\theta$ term refers to the angle-ply angle and causes e_z^0 to go from e_2^0 at $\theta = 0$ degrees to a maximum value at ± 45 degrees and then back to e_2^0 at $\theta = 90$ degrees. This means that for the case of quasi-isotropic in plane expansional strains, equation (6-18) and Example 5-7:

$$e_z^0 > e^0 = e_x^0 = e_y^0 \quad (6\text{-}26)$$

The magnitude of the extensional strains utilized in equations (6-20)–(6-25) are computed as follows:

a. Temperature change

$$e^T = \alpha \Delta T \quad (6\text{-}27)$$

and

b. Absorption of a fluid

$$e^s = \frac{1}{3}\frac{\Delta V}{V_0} \cong \frac{1}{3}\ \varrho_r M \overline{V}_D \quad (6\text{-}28)$$

$$e^s = a\ M^b \quad (6\text{-}29)$$

where ϱ_r is the density of the resin, M is the fractional weight gained in absorbing a fluid based on dry weight, $\overline{V}_D$ is the specific volume of the fluid being absorbed, V_0 is the initial dry volume of the fiber or matrix, and ΔV is the volume of fluid absorbed. The linear form of equation (6-28) assumes adativity of volumes. The current experience for bulk resins is that the volumetric change is less than the additive of volumes: therefore the power function form of equation (6-28). A typical high temperature epoxy matrix dilational stains relationships due to moisture absorption is

$$e_m^s \cong 0.42\ M^{1.15} \quad (6\text{-}30)$$

The amount of moisture absorbed when a typical epoxy is in equilibrium with the atmosphere is

$$M \cong 0.065\ (RH)^{1.5} \quad (6\text{-}31)$$

where RH is the relative humidity divided by 100 at the specified temperature, generally room temperature. Most reinforcing fiber, except polymeric based systems, permits negligible solubility of fluids, therefore

$$e_3^o \sim 0$$

Typical thermal expansion properties are selected fiber and matrix and are shown in the appendix.

Problem 6-7

It is desired to fabricate a plate possessing a zero expansion coefficient in any direction within the plane of the plate. Such a material would qualify for dimensionally stable gear applications.

Consider the problem of defining a lamination sequence such that the in-plane expansional coefficients are isotropic. A necessary and sufficient condition for this situation is that $e_1 = e_2$. Referring to equation (6-23), this occurs when the following relation is satisfied

$$\frac{R_1}{R_2} = \frac{A_{11} + A_{12}}{A_{22} + A_{12}} \tag{6-32}$$

In general, the isotropic strain induced when (6-32) is invoked is a function of the details of the lamination sequence. However, if we consider the case where $A_{11} = A_{22}$ we find from equations (6-24) and (6-32) that $H_1 = 0$ which implies that $R_1 = R_2 = J_1 h$. In this case, by use of equations (6-23) and (6-24), we obtain the following expression for the isotropic strain

$$\alpha_T = e^0 \equiv e_1^0 = e_2^0 = \frac{J_1}{U_1 + U_4} \tag{6-33}$$

Which is an *invariant* quantity. The expresssion (6-33) depends solely on the mechanical and expansional properties of a unidirectional sheet of the laminate material. We can, therefore, assert the following general principle: *for any uncoupled laminate with equal stiffnesses in two in-plane directions ($A_{11} = A_{22}$), the expansional strain field is isotropic in this plane and the normal strain in all directions is simply given by equation (6-33).* This general statement is based on the premise that all layers consist of the same material. The statement holds despite the fact that the laminate is *not* quasi-isotropic with respect to inplane stiffness. Examples of laminates which undergo isotropic expansional strain are 0°–90° bidirectional composites, ±45° angle ply, and the combined angle ply $+\alpha, -\alpha, \pi/2 + \alpha, \pi/2 - \alpha$, all of which consist of layers of equal thickness. For a given composite material, all of these systems undergo identical expansional strains.

Equation (6-33) may also be expressed in terms of engineering constants as

$$e^0 = W_1 + \frac{2(E_{11} - E_{22})\, W_2}{E_{11} + (1 + 2\nu_{12})\, E_{22}} \tag{6-34}$$

where all moduli refer to the properties of the unidirectional material and W_1,

W_2 are given in equations (6-24). According to equation (6-34) as the ratio $E_{11}/E_{22} \rightarrow \infty$, the isotropic strain becomes

$$e^0 = W_1 + 2W_2 = e^{0L}_1 \; as \; \frac{E_{11}}{E_{22}} \rightarrow \infty \qquad (6\text{-}35)$$

Thus for very highly anisotropic composites (high ratio of E_{11}/E_{22}), the isotropic strain approaches the expansional strain of a unidirectional ply parallel to the filaments.

The corresponding thickness strain, equation (6-25), e_z when the ratio of E_{11} to E_{22} is large is

$$e^0_z \cong e^0_2 \, (1 + \nu_{23})$$

This situation was observed when the nylon-rubber material, Figure 6-14, was laminated to generate an isotropic in-plane expansional strain field in which

$$e^0_x = e^0_y \sim 0$$

and

$$e^0_z \sim 1.2$$

Small or zero in-plane expansion fields are therefore obtained at the expense of large thickness direction deformations which complicate structural integrity issues.

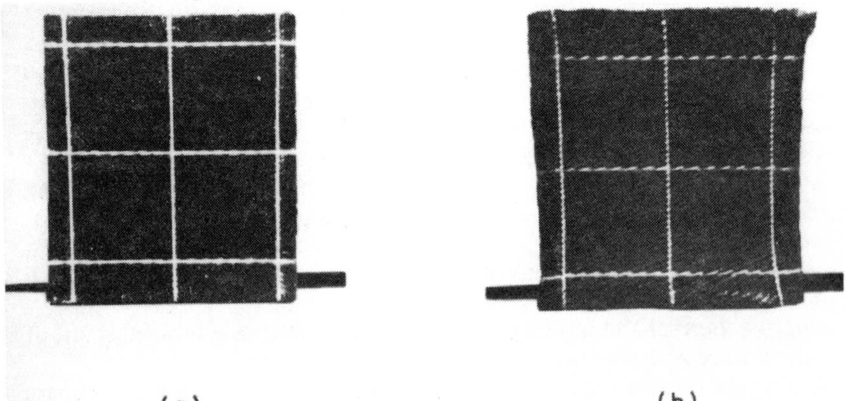

(a) (b)

Figure 6-14. Comparison of a dry and swollen $(0/90)_s$ laminate of nylon elastomeric ply material.

Problem 6-8

It is desired to a laminated plate or "board" to support electronic components which will be soldered to the surface of the laminate. At these solder locations the laminated board will be penetrated with small holes which are plated with copper to conduct electrical current. Both the solder and copper will fail mechanically if the induced expansion strains in the laminate are sufficient. The solder joints will be sensitive to the magnitude of e_1^0 and e_2^0. The plated through hole will be sensitive to e_3^0. Using the data of J.R. Strife and K.M. Prewo, *J. Comp. Mat., 13,* 264 (1979) for a Kevlar/PR-286 material system:

$$v_f = 0.5$$

$E_{11} = 9.3 \times 10^{-6}$ psi	$G_{12} = 0.3 \times 10^{-6}$ psi
$E_{22} = 7.8 \times 10^{-6}$ psi	$\nu_{12} = 0.029$
$\alpha_1 = 1.2 \times 10^{-6}/°F$	$\alpha_2 = 40.0 \times 10^{-6}/°F$

a. Compute the in-plane expansion for bidirectional laminates of $\pm\theta$ equal to 22°, 30°, 45°, and 65°. Compared with the measured properties of

$\pm\theta$	$\alpha_1 \times 10^{-6}/°F$
22°	-6.9
30°	-7.6
45°	1.5
60°	27.6
68°	37.8

b. Compute the "out-of-plane" expansion coefficient, α_3 for the same bidirectional laminates using equations (6-25) and (6-15). Is α_3 a maximum when $\alpha_1 = \alpha_2$?

c. In an environment in which the expansional strains will be large and variable in time at what combinations of $\pm\theta$ would produce the least interlaminar stress, see (4-37) and (4-30) when $e_1^0 = e_2^0$?

d. Compute the laminate expansional strains e_1^0, e_2^0, and e_3^0 for the Kevlar/PR-286 material when the material is in equilibrium with 75 percent relative humidity (~4 percent absorbed moisture by weight).

e. Compare the magnitude of the thermally induced strains $e_1^0 = e_2^0$ and e_3^0, items a and b, with the moisture induced strains, item d. Use a 250°F operating temperature range for the comparison.

f. Sum the thermal and moisture strains to appreciate the combined effort of temperature and moisture.

g. Repeat the computations for a 0/90 S-glass/epoxy material and compare with the Kevlar material. What are the differences? Which one would you choose for a strain limited design? Why?

REFERENCES

1. Chamis, C. C. and Sendeckyj, G. P., "Critique on Theories Predicting Thermoelastic Properties of Fibrous Composites," *J. of Comp. Mat., 2,* 332 (1968).
2. Hill, R., "Theory of Mechanical Properties of Fibre-Strengthened Materials: III. Self Consistent Model," *J. Mech. Phys. Solids, 13,* 189 (1965).
3. Kilchinskii, A. A., "On One Model for Determining Thermoelastic Characteristics of Fiber Reinforced Materials," (in Russian), *Prikladnaia Mekhanika, 1*(12), 65 (1965).
4. Kilchinskii, A. A., "Approximate Method for Determining the Relation between the Stresses and Strains for Reinforced Materials of the Fiber Glass Type," (in Russian), *Thermal Stresses in Elements of Construction,* Vol. 6. Kiev: Naukova Dumka, p. 123 (1966).
5. Hermans, J. J., "The Elastic Properties of Fiber Reinforced Materials when the Fibers are Aligned," *Proc. Konigl. Nederl. Akad. van Weteschappen Amsterdam, B70*(1), 1 (1967).
6. Hershey, A. V., "The Elasticity of an Isotropic Aggregate of Anisotropic Cubic Crystals," *J. Appl. Mech., 21,* 236 (1954).
7. Kroner, E., "Berechnung der Elastichen Konstanteu des Vielkristalls aus den Konstanten des Einkristalls," *Z. Physik, 151,* 504 (1958).
8. Eshelby, J. D., "Elastic Inclusions and Inhomogeneities," in *Progress in Solid Mechanics,* Vol. 2, ed. Sneddon & Hill. Amsterdam: North-Holland, p. 89 (1961).
9. Frohlich, H. and Sack, B., "Theory of the Rheological Properties of Dispersions," *Proc. Roy. Soc. Lond., A185,* 415 (1946).
10. Ebert L. J., Hamilton, C. H., and Hecker, S. S., "Development of Design Criteria for Composite Materials," *AFML TR* 67-95 (April 1967).
11. Whitney, J. M. and Riley, M. B., "Elastic Properties of Fiber Reinforced Composite Materials," *J. AIAA, 4,* 1537 (1966).
12. Whitney, J. M., "Elastic Moduli of Unidirectional Composites with Anisotropic Filaments," *J. Composite Materials, 1,* 188 (1967).
13. Whitney, J. M., "Geometric Effects of Filament Twist on the Modulus and Strength of Graphite Fiber-Reinforced Composites," *Textile Res. J., 36,* 765 (1966).
14. Paul, B., "Prediction of Elastic Constants of Multiple Materials," *Trans. of the Metall. Society of AIME, 218,* 36 (1960).
15. Hashin, Z. and Rosen, B. W., "The Elastic Moduli of Fiber Reinforced Materials," *J. Appl. Mechanics, Trans. of the ASME, 31,* 233 (1964) (Errata—*J. Appl. Mech., 32,* 219 (1965), also NASA CR-31 (1964)).
16. Abolin'sh, D. S., "Compliance Tensor for an Elastic Material Reinforced in One Direction," *Polymer Mechanics, 1*(4), 28 (1965).
17. Dow, N. F. and Rosen, B. W., "Evaluation of Filament-Reinforced Composites for Aerospace Structural Applications," NASA CR-207 (April 1965).
18. Hashin, Z., "On Elastic Behavior of Fiber-Reinforced Materials of Arbitrary Transverse Phase Geometry," *J. Mech. Phys. Solids, 13,* 119 (1965).
19. Hill, R., "Theory of Mechanical Properties of Fiber-Strengthened Materials: I. Elastic Behavior," *J. Mech. Phys. Solids, 12,* 199 (1964).
20. Levin, V. M., "On the Coefficients of Thermal Expansion of Heterogeneous Materials," (in Russian), *Mekhanika Tverdovo Tela* (1), 88 (1967).
21. Schapery, R. A., "Thermal Expansion Coefficients of Composite Materials Based on Energy Principles," *J. Comp. Mat., 2,* 380 (1968).
22. Kurshin, L. M. and Fil'shtinskii, L. A., "Determination of Reduced Elastic Moduli of an Isotropic Plane, Weakened by a Doubly-Periodic Array of Circular Holes," (in Russian), *Izv. Akad, Nauk SSSR, Mekh. Mash.* (6), 110 (1961).
23. Fil'shtinskii, L. A., "Stresses and Displacements in an Elastic Sheet Weakened by a Doubly-Periodic Set of Equal Circular Holes," *P.M.M., 28,* 430 (1964).
24. Grigolyuk, E. I., Kurshin, L. M., and Fil'shtinskii, L. S., "On One Method of Solving the Doubly-Periodic Problem of Theory of Elasticity," (in Russian), *Prikladnaia Mekhanika, 1*(1), 22 (1965).
25. Grigolyuk, E. I. and Fil'shtinskii, L. A., "Elastic Equilibrium of an Isotropic Plane with a Doubly-Periodic Set of Inclusions," (in Russian), *Prikladnaia Mekhanika, 2*(9), 1 (1966).

26. Van Fo Fy, G. A., "On the Equations Connecting the Stresses and Strains in Glass Reinforced Plastics," (in Russian), *Prikladnaia Mekhanika, 1*(2), 110 (1965).
27. Van Fo Fy, G. A., "Stressed and Deformed State of Synthetic Materials in Shear," (in Russian), *Prikladnaia Mekhanika, 1*(5), 111 (1965).
28. Van Fo Fy, G. A., "On the Theory of Anisotropic Creep of Unidirectional Glass-Reinforced Plastic," *Mekhanika Polimerov, 1*(2), 65 (1965); Van Fo Fy, G. A., "Elastic Constants and Stressed State of Glass-Fiber Ribbon," (in Russian), *Mekhanika Polimerov* (4), 593 (1966).
29. Van Fo Fy, G. A. and Klyavlin, V. F., "Investigation of the Dependence of Mechanical Properties and Internal Stress Fields in Shear on the Microstructure of Reinforced Media," (in Russian), *Mekhanika Polimerov* (4), 667 (1967).
30. Van Fo Fy, G. A. and Savin, G. N., "Fundamentals of the Theory of Non-Fabric Glass-Reinforced Plastics," *Mekhanika Polimerov, 1*(1), 151 (1965); Van Fo Fy, G. A., "Thermal Strains and Stresses in Glass Fiber Reinforced Media," (in Russian), *Prikl. Mekh. i Teor. Fix.* (4), 101 (1965); Van Fo Fy, G. A., "Elastic Constants and Thermal Expansion of Certain Bodies of Inhomogeneous Regular Structure," *Soviet Physics—Doklady, II*(2), 176 (1966) (Translation of Doklady Akad. Nauk SSSR, *166*(4), 817 (1965)); Van Fo Fy, G. A., "On the Basis of Theory of Anisotropic Thermoviscoelasticity," (in Russian), in *Thermal Stresses in Elements of Construction.* Vol. 6, Kiev: Naukova Dumka, p. 109 (1966).
31. Van Fo Fy, G. A., "Basic Relations of Theory of Oriented Glass-Reinforced Plastics with Hollow Fibers," (in Russian), *Mekhanika Polimerov* (5), 763 (1966); Van Fo Fy, G. A., "Initially Stressed Oriented Glass-Reinforced Plastics with Hollow Fibers," (in Russian), *Prikladnaia Mekhanika, 2*(7), 1 (1966).
32. Van Fo Fy, G. A., "Diffusion in Reinforced Bodies," (in Ukranian), *Dopovidi Akad. Nauk UkSSR, Series A*(10), 891 (1967).
33. Van Fo Fy, G. A., "Investigation of the Effect of Treatment of Fibers on the Stress Distribution in the Structure of Glass-Reinforced Plastics (in Russian), *Prikladnaia Mekhanika, 3*(2), 106 (1967).
34. Herrmann, L. R. and Pister, K. S., "Composite Properties of Filament-Resin Systems," ASME P.N. 63 WA-239, Paper presented at the ASME Annual Meeting, Philadelphia, Pennsylvania (November 17-22, 1963).
35. Wilson, H. B., Jr. and Hill, J. L., "Mathematical Studies of Composite Materials," Rohm and Haas Special Report No. S-50, AD 468 569 (1965).
36. Adams, D. F., Doner, D. R., and Thomas, R. L., "Mechanical Behavior of Fiber-Reinforced Composite Materials," Wright-Patterson Air Force Base, AFML TR 67-96, AD 654065 (May 1967).
37. Adams, D. F. and Doner, D. R., "Longitudinal Shear Loading of a Unidirectional Composite," *J. Composite Materials, 1,* 4 (1967).
38. Adams, D. F. and Doner, D. R., "Transverse Normal Loading of a Unidirectional Composite," *J. Composite Materials, 1,* 152 (1967).
39. Pickett, G., "Analytical Procedures for Predicting the Mechanical Properties of Fiber Reinforced Composites," AFML TR 65-220, Also AD 473790 (1965).
40. Pickett, G. and Johnson, M. W., "Analytical Procedures for Predicting the Mechanical Properties of Fiber Reinforced Composites," AFML TR 65-220, Pt. 2, Also AD 646 216 (1966).
41. Chen, C. H. and Cheng. S., "Mechanical Properties of Fiber Reinforced Composites," *J. Composite Materials, 1,* 30 (1967).
42. Clausen, W. E. and Leissa, A. W., "Stress and Deflection Analysis of Fibrous Composite Materials under External Load," AFML TR 67-15 AD 656 431 (1967).
43. Bloom, J. M. and Wilson, H. B., "Axial Loading of a Unidirectional Composite," *J. Composite Materials, 1,* 268 (1967).
44. Quackenbush, N. E. and Thomas, R. L., "Investigation of Carbon Filament Reinforced Plastics," Philco-Ford, AD 820 492 (1967).
45. Foye, R. L., "An Evaluation of Various Engineering Estimates of the Transverse Properties of Unidirectional Composites," *SAMPE, 10,* G-31 (1966).
46. Foye, R. L., "Structural Composites," Quarterly Progress Reports No. 1 and 2, AFML Contract No. AF 33(615)-5150 (1966).
47. Grigolyuk, E. I. and Fil'shtinskii, L. A., "Perforated Plates and Shells, and Related Problems, Survey of Results," in *Uprugost i Plastichnost' 1965,* 7 (Moscov, 1967).

48. Lomakin, V. A. and Koltunov, M. A., "Effect of Reinforced Elements on the Deformation and Strength of Glass-Reinforced Plastics in Tension," (translation of Mekhanika Polimerov) *Polymer Mechanics, 1*(2), 79 (1965).
49. Haener, J. and Ashbaugh, N., "Three Dimensional Stress Distribution in a Unidirectional Composite," *J. Composite Materials, 1,* 54 (1967) (Also AD 643 813).
50. Piehler, H. R., "Interior Elastic Stress Field in a Continuous Close-packed Filamentary Composite Material under Uniaxial Tension," *Fiber-Strengthened Metallic Composites,* ASTM, STP 427 Am. So. Testing Mats., 3 (1967).
51. Meijers, P., "Doubly-Periodic Stress Distributions in Perforated Plates," Ph.D. Dissertation, Technische Hogeschool, Delft (1967).
52. Behrens, E., "Elastic Constants of Filamentary Composites with Rectangular Symmetry," *J. Acoust. Soc. Am., 42,* 367 (1967).
53. Behrens, E., "Thermal Conductivities of Composite Materials," *J. Composite Materials, 2,* 2 (1968).
54. Springer, S. G. and Tsai, S. W., "Thermal Conductivities of Unidirectional Materials," *J. Composite Materials, 1,* 166 (1967).
55. Vishneskii, G. E. and Shlenskii, O. F., "Effect of Component Properties and the Geometric Characteristics of the Structure on the Values of Coefficients of Heat Conduction in Fiber Reinforced Plastics," in Russian, *Mekhanika Polimerov, 4*(1), 18 (1968).
56. Halpin, J. C. and Kardos, J. L., *Polym. Eng. and Sci., 16,* 344 (1976); Halpin, J. C. and Tsai, S. W., "Environmental Factors in Composite Materials Design," AFML TR 67-423.
57. Hewitt, R. L. and deMalherbe, M. C., *J. Comp. Mat., 4,* 280 (1970).
58. Halpin, J. C., *J. Comp. Mat., 3,* 732 (1969).
59. Halpin, J. C. and Pagano, N. J., *J. Comp. Mat., 3,* 720 (1969).
60. Piehler, H. R., "Interior Elastic Stress Field in a Continuous Close-packed Filamentary Composite Material under Uniaxial Tension," *Fiber-Strengthened Metallic Composites,* ASTM, STP 427 Am. So. Testing Mats., 3 (1967).
61. Yeow, Y. T., *J. Comp. Mat.,* Supplement, 132 (1980).
62. Halpin, J. C., Terina, K., and Whitney, J. M., *J. Comp. Mat., 5,* 36 (1971).
63. Halpin, J. C. and Kardos, J. L., *J. Apply. Phys., 43,* 2235 (1972).
64. Nicolais, L., *Poly. Eng. and Sci., 15,* 137 (1975).
65. Halpin, J. C. and Pagano, N. J., "Consequences of Environmentally Induced Dilation in Solids" in *Proc. 6th Annual Mtg., Soc. of Eng. Sci.,* New York: Springer-Verlag (1969).
66. Springer, S. G. and Tsai, S. W., "Thermal Conductivities of Unidirectional Materials," *J. Composite Materials, 1,* 166 (1967).
67. Gogol, W. and Furmanski, P., *J. Comp. Mat.,* Supplement, 167 (1980).
68. Z. Hashin, *J. Comp. Mat., 2,* 284 (1968).
69. DeIsai, R. and Whiteside, J. B., in *Advanced Composite Materials—Environmental Effects.* ASTM STP 658, p. 2 (1978).
70. Loos, A. C. and Springer, G. S., *J. Comp. Mat., 13,* 131 (1979).
71. Halpin, J. C. and Thomas, R. L., *J. Comp. Mat., 2,* 488 (1968).
72. Fahmy, A. A. and Ragai-Ellozy, A. N., *J. Comp. Mat., 8,* 90 (1974).
73. Pagano, N. J., *J. Comp. Mat., 8,* 310 (1974).

7

Characterization and Behavior of Structural Composites

7.1 INTRODUCTION

CHARACTERIZATION PROCEDURES FOR ANISOTROPIC COMPOSITE materials must provide not only experimental data for moduli or stiffness properties, but also must characterize the complete mechanical response of the material up to and including the ultimate failure of the system. The analytical and experimental results presented here clearly demonstrate that many classical notions associated with mechanical characterization experiments, extended from the technology of isotropic solids, must be revised in connection with the testing of anisotropic solids. The failure of current accepted test procedures necessitates new and unique testing procedures for both unidirectional and multidirectional composites.

7.2 STIFFNESS MEASUREMENTS

The constitutive equations with respect to the material symmetry axes of a unidirectional composite sheet are given by

$$\varepsilon_1 = S_{11}\sigma_1 + S_{12}\sigma_2 \tag{7-1}$$

$$\varepsilon_2 = S_{12}\sigma_1 + S_{22}\sigma_2 \tag{7-2}$$

$$\varepsilon_6 = S_{66}\sigma_6 = \gamma_{12} = S_{66}\tau_{12} \tag{7-3}$$

for a state of plane stress in the plane of the sheet. In these equations the subscripts 1, 2 refer to normal stress and strain, 6 indicates shear components (e_6 is the *engineering* shear strain), and

$$S_{11} = \frac{1}{E_{11}}, \; S_{22} = \frac{1}{E_{22}}, \; S_{12} = -\frac{\nu_{12}}{E_{11}}, \; S_{66} = \frac{1}{G_{12}} \tag{7-4}$$

For isotropic materials, only two of these material coefficients are independent. These cases are illustrated in Figures 7-1a and 7-1b.

For unidirectional composite materials in which the fibers are oriented at an arbitrary angle to the edges of the sheet (Figure 7-1c), the constitutive equations take the form

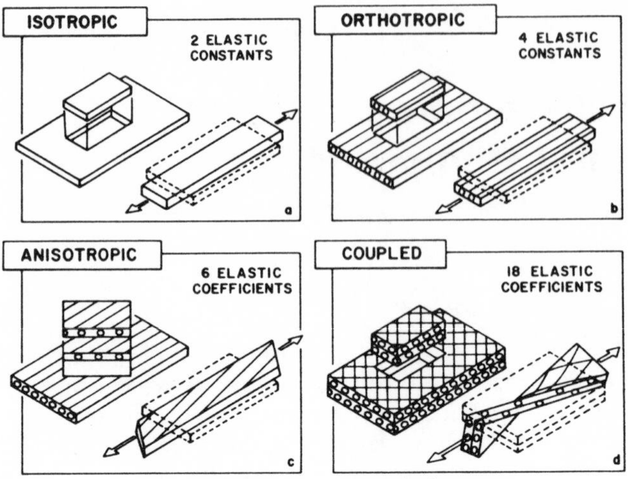

Figure 7-1. *Composite terminology.*

$$\varepsilon_1' = S_{11}'\sigma_1' + S_{12}'\sigma_2' + S_{16}'\sigma_6' \tag{7-5}$$

$$\varepsilon_2' = S_{12}'\sigma_1' + S_{22}'\sigma_2' + S_{26}'\sigma_6' \tag{7-6}$$

$$\varepsilon_6' = S_{16}'\sigma_1' + S_{26}'\sigma_2' + S_{66}'\sigma_6' \tag{7-7}$$

where primed quantities are measured with respect to axes parallel to the edges of the sheet and

$$S_{16}' = -\frac{\eta_{16}'}{E_{11}'}, \quad S_{26}' = -\frac{\eta_{26}'}{E_{22}'} \tag{7-8}$$

By use of the standard transformation equations [7-1], S_{ij}' can be expressed in terms of S_{ij} and fiber orientation. As a consequence of the shear coupling compliances S_{16}' and S_{26}', the response of an "off-angle" composite (i.e., a specimen in which the filaments are neither parallel nor perpendicular to the direction of the applied force) under uniaxial tension appears as shown in Figure 7-1c.

For laminated materials, the elastic properties of the individual layers are expressed in terms of the reduced matrix Q_{ij} and from classical laminated plate theory [7-1], the laminate constitutive relations are expressed as

$$\begin{bmatrix} N_1 \\ N_2 \\ N_6 \\ M_1 \\ M_2 \\ M_6 \end{bmatrix} = \begin{bmatrix} A_{11} & A_{12} & A_{16} & B_{11} & B_{12} & B_{16} \\ A_{12} & A_{22} & A_{22} & B_{12} & B_{22} & B_{26} \\ A_{16} & A_{26} & A_{66} & B_{16} & B_{26} & B_{66} \\ B_{11} & B_{12} & B_{16} & D_{11} & D_{12} & D_{16} \\ B_{12} & B_{22} & B_{26} & D_{12} & D_{22} & D_{26} \\ B_{16} & B_{26} & B_{66} & D_{16} & D_{26} & D_{66} \end{bmatrix} \begin{bmatrix} \varepsilon_1 \\ \varepsilon_2 \\ \varepsilon_6 \\ \varkappa_1 \\ \varkappa_2 \\ \varkappa_6 \end{bmatrix} \tag{7-9}$$

where

$$(A_{ij}, B_{ij}, D_{ij}) = \int_{-h/2}^{h/2} Q_{ij}(1,z,z^2)dz \tag{7-10}$$

$$(N_i, M_i) = \int_{-h/2}^{h/2} \sigma_i(1,z)dz, \; i,j = 1,2,6 \tag{7-11}$$

As pointed out in Chapter 3 equation (7-9) reveals a coupling phenomenon between bending, twisting, and extension, e.g., a two layer angle-ply composite under uniaxial tension twists as shown in Figure 7-1d. In the event that Q_{ij} is an even function of z (symmetric layup), B_{ij} vanishes and the constitutive equations become uncoupled.

The principal problems encountered in experimental characterization arise from the coupling coefficients S_{16}, S_{26}, Q_{16}, Q_{26}, D_{16}, D_{26}, and B_{ij} in the constitutive equations. The lack of appreciation of the physical consequences of these coupling terms has resulted in the current deficiencies in characterization and analysis. The concept phenomena is peculiar to anisotropic and layered systems and it is for this reason that the technology developed for isotropic solids must be modified for anisotropic bodies.

7.3 FRACTURE CHARACTERIZATION

The characterization of the fracture properties of anisotropic solids involves the development of a strength criterion as well as fracture mechanics or toughness and fatigue characterization. Knowledge of the state of stress, which is preferably uniform, particularly when material response is nonlinear, in strength determinations is absolutely essential if we are to have a large base of experimental evidence upon which strength and fatigue theories may be deduced. The development of a failure criterion for fiber reinforced materials involves a great amount of off-angle testing of unidirectional composites in order to establish transformation properties of "strength." Furthermore, while it is generally recognized that strength properties in tension and compression are functions of fiber orientation, it is not generally appreciated that the shear strength of an anisotropic material is strongly dependent on the direction of the shear stress [7-2]. For example, consider the unidirectional 45° composite shown in Figure 7-2 under states of pure shear of opposite sense. Owing to the directional properties of the material, the nature of the fracture, as well as the ultimate value of τ, in these two cases will be vastly different. For the state of stress in Figure 7-1a, the fracture surface would tend to align parallel to the filaments, while in Figure 7-1b, the filaments themselves would be fractured. This feature of anisotropic material response has recently been demonstrated experimentally by Halpin and Wu [7-3] (see Table 7-1).

The prediction of the lifetime and failure modes for damaged anisotropic bodies involves an experimental fracture mechanics program. The goal of a program of this nature is to specify the applied stresses (or loads) and geometry (crack length, specimen dimensions) to simulate in the laboratory

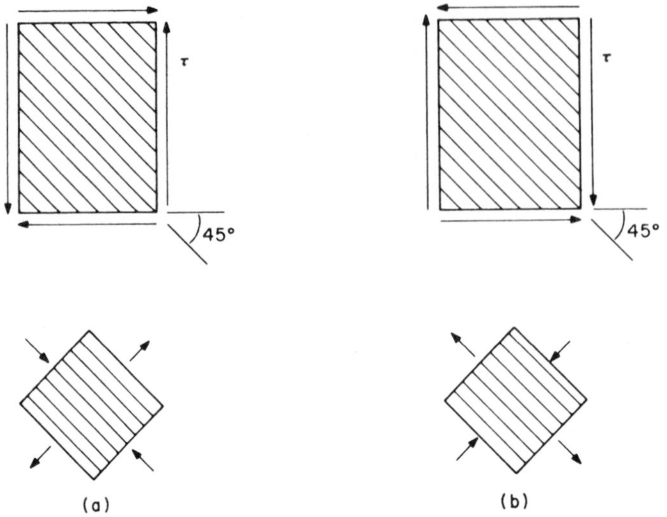

Figure 7-2. Illustration of shear stress directions on strength.

the conditions which will cause crack extension and failure of a structure in service. A typical experiment in this area, similar to that performed by Wu [7-4], studies crack extension under (a) tension and shear, (b) pure shear, and (c) compression as shown in Figure 7-3.

7.4 ANALYSIS OF CHARACTERIZATION PROCEDURES

One of the most elementary concepts in elasticity theory is that of a uniform state of stress. Producing such a state of stress in the characterization

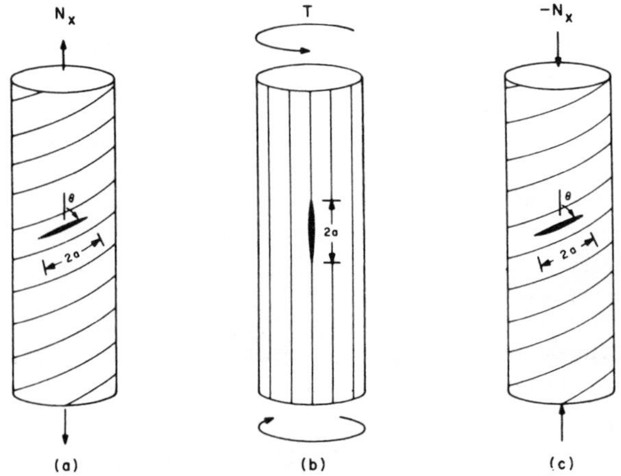

Figure 7-3. Fracture mechanics experiment on fibrous composite.

experiment is absolutely essential to determine fundamental material properties; however, this is no trivial task. For example, the common tension test of off-angle specimens can yield strength and stiffness data which is greatly erroneous [7-5,6]. Suppose that the tension test of an off-angle composite is used as the basis for determining E'_{11}, the composite modulus of elasticity parallel to the long edges of the specimen. If the effects of end constraint are not taken into account, this modulus will be erroneously recorded as E_{11}^*, where [7-5]

$$E_{11}^* = \frac{1}{S'_{11}}\left(\frac{1}{1-\eta}\right) \tag{7-12}$$

with

$$E'_{11} = \frac{1}{S'_{11}} \tag{7-13}$$

and

$$\eta = \frac{6S'^{\,2}_{16}}{S'_{11}\left(6S'_{66} + S'_{11}\,\dfrac{\ell^2}{h^2}\right)} \tag{7-14}$$

The term involving η represents a conservative estimate of the error in this experiment. The error results because conventional clamping devices induce severe perturbations (bending and shear) in the stress field [7-5,7]. These large effects are a direct consequence of the shear coupling compliance, S'_{16}. The response of an off-angle composite under uniform tension is shown in Figure 7-4a, while the corresponding clamped specimen is depicted in Figure 7-4b. Uniform states of stress and strain are only possible when $S'_{16} = 0$. This condition, of course, is satisfied for 0° or 90° fiber orientation so that these specimens can be utilized to evaluate the compliance coefficients S_{11}, S_{12}, and S_{22} and the analogous engineering constants and strength properties. The shear compliance S_{66} must be determined from an additional experiment.

An acceptable experiment for defining the shear modulus G_{12} and interlaminar shear stress (at least to the point of material non-linearity) is the pure torsion of a solid circular cylindrical rod with the fibers oriented parallel to the longitudinal axis. The analysis of this specimen is identical to that of an isotropic cylinder under torsion, *T*. This test configuration has been employed by Adams, Doner, and Thomas [7-13] among others for the determination of unidirectional composite shear properties. The maximum stress and shear modulus can be computed from the classical strength of materials relations

$$\tau_{max} = \frac{2T}{\pi R^3} \tag{7-15}$$

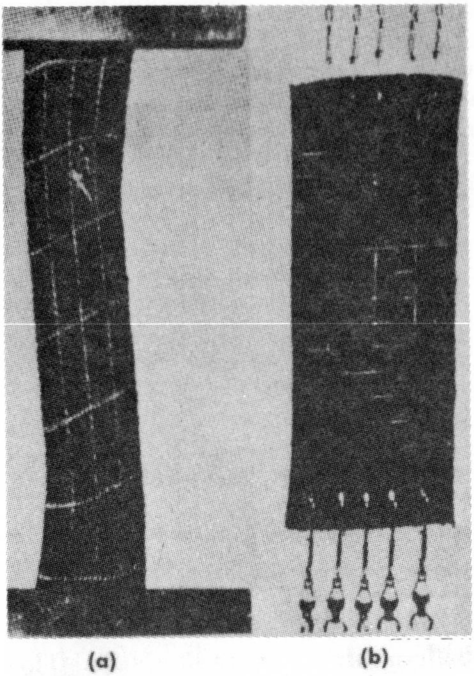

Figure 7-4. *Improper and proper unaxial experiment.*

$$G_{12} = \frac{2T}{\pi R^3 \varepsilon_6} \tag{7-16}$$

where ε_6 is the maximum shear strain on the surface of the rod.

Next to tension testing the most common experiment employed to define the response of composite materials is the flexure test. Although this test is generally invalid for strength determination owing to material non-linearity and stress concentrations in the region of an applied force, there are instances where flexure experiments can yield useful information if the data is properly interpreted.

In 1949, Hoff [7-8] presented flexure formulas for laminates, and recently Pagano [7-9] has modified these expressions to account for shear deflection. Although these formulas are based upon elementary strength of materials considerations, they appear quite adequate to describe the special cases of deflection of unidirectional and bidirectional (0°–90°) beams. While the neglect of shear deflection is long isotropic beams is a valid assumption, shear deflection in composite beams can be very significant due to the low ratio of shear modulus to longitudinal modulus for most composites.

Despite the connection for shear deflection however, the flexure analysis is currently incomplete since it cannot describe the complex response of a laminated beam of arbitrary layer orientations and stacking sequence.

We now consider the deflection of a unidirectional (off-angle) beam induced by pure bending moments. Two cases are of interest to us: (1) the fibers are parallel to the front face of the members as in Figure 7-5a, and (2) the fibers are parallel to the top surface as shown in Figure 7-5b.

For case (1), Figures 7-5a and 7-6a, it can be shown [7-10] that the center line vertical displacement, v, is independent of S_{16} and corresponds to that given by the elementary Euler-Bernoulli beam theory. If the ends, $x = o, \ell$, are pinned, the deflection surface is given by

$$v = \frac{MS_{11}}{2I}(\ell x - x^2) \tag{7-17}$$

where it is understood that the compliance S_{11} is the primed quantity given in equation (7-5). It should be noted however that the Euler-Bernoulli hypothesis itself is not satisfied since the horizontal displacement u is a non-linear function of y (u also depends upon S_{16}).

Consider now the orientation of case (2), Figures 7-4b and 7-5b, and let the vertical central-plane displacement be represented by w. The three dimensional elasticity solution for the maximum deflection w_0 is given as

$$w_0 = \frac{3M_0 b^2}{2h^3}(S_{11}C^2 + S_{16}C + S_{12}) \tag{7-18}$$

when C is the length-to-width ratio and b is the beam width. Equation (7-18) is based upon certain assumed boundary conditions on w. Although these boundary conditions are plausible, it is quite likely that the boundary conditions in a physical experiment are impossible to define owing to the warping exhibited by the beam as shown in Figure 7-6b. The term involving S_{16} in-

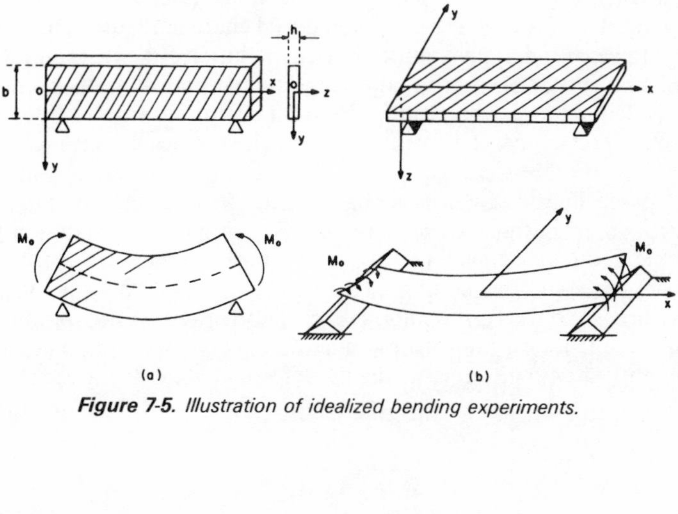

Figure 7-5. *Illustration of idealized bending experiments.*

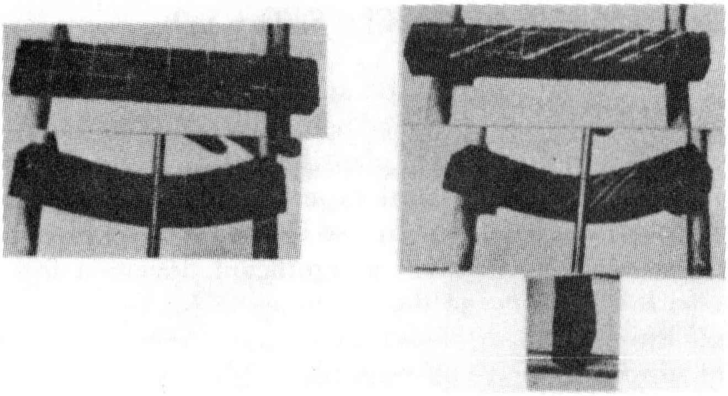

(a) (b)

Figure 7-6. Experimental bending observations.

volves a significant deviation from classical beam theory; it is because of the existence of S_{16} that the beam twists and lifts off the support as shown in Figures 7-5b and 7-6b. If one attempts to suppress the lift-off from the supports by employing double knife edge supports, twisting moments will be induced in the member. As the value of C increases, the influence of S_{12} diminishes rapidly, while the effect of S_{16} dissipates only for very large values of C. In fact the apparent flexure modulus of a unidirectional 10° off-angle specimen is higher than that of a 0° specimen for the material parameters considered in [7-12].

Although we have only discussed pure bending here, we can expect that beams loaded by concentrated forces would display at least qualitatively, similar response characteristics. Unfortunately, the exact elasticity solutions for the configurations of cases (1) and (2) under concentrated forces do not exist. This fact, together with the other objections noted above, lead us to reject these modes of testing for precise material characterization, in particular, case (2). However, an approximate elasticity solution for case (1) under a central load may be derived from the cantilever beam solution presented by Hashin [7-11]. Hence, this configuration may merit some attention as a means of *stiffness* characterization if specimen stability does not prove to be critical.

In the preceeding discussion we have demonstrated some of the practical complications resulting from the unique response characteristics of anisotropic and layered media. If we are required to measure shear modulus or produce a well-defined combined state of stress, or even a state of uniaxial stress, we are led to the consideration of yet another test configuration, namely the thin-walled hollow anisotropic cylinder. This specimen, i.e., a helical wound composite tube, serves as the basis for a general characterization procedure (see Figure 7-3). Again the primary concern is the existence of a uniform state of stress under practical testing conditions. While the elasticity analysis shows that a uniform state of stress cannot exist in such a specimen unless S'_{16} vanishes [7-14], the stress field is nearly uniform for thin tubes. An acceptable characterization procedure for uniaxial tension or combined loadings can be achieved in the laboratory by allowing for the unconstrained rotation and longitudinal motion of one end of the tube [7-14]. The rotation of the helical wound tube under tension or compression is a natural consequence of the shear coupling compliance S'_{16}. Conversely, the state of pure shear induced by torsion results in either an increase or decrease in the height of the cylindrical tube. If these motions are prohibited, extraneous stresses are induced in the specimen. Photographs of deformed helical wound tubes are under uniform tension and internal pressure shown in Figures 7-7a, 7-7b, respectively.

The next step is the analysis of layered shells under combined loading conditions. Using the well-known Donnell approximations, Chapter 4, employed by Dong, Pister, and Taylor [7-15] in their analysis of thin laminated anisotropic cylindrical shells, a relatively straightforward solution, Chapter 5, can be obtained [7-16] under combined tension (or compression), shear, and internal pressure. In this study by Whitney and Halpin, all six components of the in-plane stiffness matrix are defined in terms of experimentally observable quantities for arbitrary layer orientations.

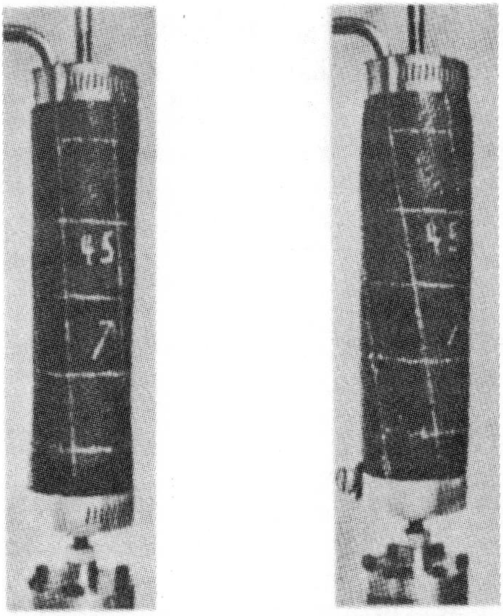

(A) (B)

Figure 7-7. *Tension (A) and internal pressure (B) in an anisotropic thin-walled tube.*

$$e_x^0 = (A_{11}^* + A_{16}^* K_1)N_x + (A_{12}^* + A_{16}^* K_2)Rp + A_{16}^* K_3 T \quad (7\text{-}19)$$

$$e_y^0 = (A_{12}^* + A_{26}^* K_1)N_x + (A_{22}^* + A_{26}^* K_2)Rp + A_{26}^* K_3 T \quad (7\text{-}20)$$

$$e_{xy}^0 = (A_{16}^* + A_{66}^* K_1)N_x + (A_{26}^* + A_{66}^* K_2)Rp + A_{66}^* K_3 T \quad (7\text{-}21)$$

where

$$K_1 = \frac{B_{16}^*}{R}, \; K_2 = \frac{B_{26}^*}{R}, \; K_3 = \frac{1}{2\pi R^2} \quad (7\text{-}22)$$

and

$$A^* = A^{-1}$$
$$B^* = -A^{-1}B \quad (7\text{-}23)$$

For symmetric layups $(B_{ij} = 0)$

$$K_1 = K_2 = 0 \quad (7\text{-}24)$$

The quantities R, p, T, and N_x are the radius, internal pressure, applied torque, and longitudinal normal stress resultant, respectively. For a symmetric composite, equations (7-19, 24) reveal, as discussed in earlier chapters, that all components of A_{ij}^* can be determined by applying combinations of axial force, internal pressure, and torsion to the sample. As a consequence, all of the equivalent engineering constants can be evaluated in this series of experiments. For example, consider the shear determination for a thin-walled orthotropic or symmetric tube, equations (7-19)–(7-24):
From equation (7-21)

$$\varepsilon_{xy} = \frac{A_{66} T}{2\pi R^2}$$

The angular rotation, θ, of a tube length, l, subjected to a torque is:

$$\theta = \frac{\varepsilon_{xy} l}{R}$$

and

$$A_{66}^* \equiv \frac{S_{16}}{h} = \frac{1}{G_{12} h}$$

Therefore,

$$G_{12} = \frac{4Tl}{G_{12}\pi h D^3} \tag{7-25}$$

The corresponding stress averaged across the thickness is

$$\bar{\sigma}_{xy} = \frac{T}{2h\pi R^2} \tag{7-26}$$

However, for detailed analysis it must be recognized that the stress varies from layer to layer in a laminate Chapter 5. It should be noted, however, that certain difficulties arise in defining the engineering constants of laminated materials, as pointed out by Whitney [7-17]. These constants can always be expressed in terms of the in-plane compliance matrix, but not in terms of the in-plane stiffness matrix alone unless the laminate is balanced and A_{ij} is orthotropic. This fact often leads to confusion in the reporting of experimental data. For example, owing to the variation of stresses through the thickness, the shear modulus of a $\pm 45°$ symmetric angle-ply composite is not equal to the shear modulus of a 45° unidirectional layer, even though the latter shear modulus is uniform through the thickness.

The response of a laminate which is not symmetrical through the thickness (a two-layer angle-ply) under uniaxial tension is illustrated in Figure 7-1d and shown experimentally in Chapter 3 and again in Figure 7-8. Despite the fact

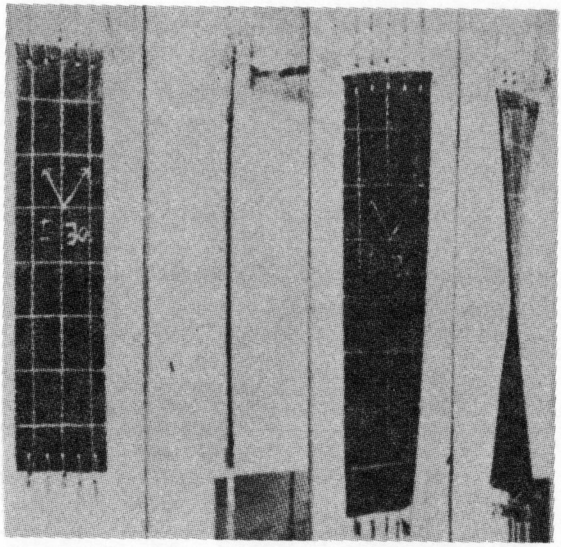

(a) (b)

Figure 7-8. *B_{16} coupling between bending and stretching: a) unsymmetrical angle-ply in rest state; and b) in a state of unaxial tension.*

that the fiber orientations are symmetric with respect to the load direction, coupling between twisting and extension exists owing to B_{16}, equation (7-9). Obviously, the constraints exerted by conventional clamping devices on such a specimen will induce a highly complex state of stress in the sample. Similar phenomena have been observed in flexure experiments for *symmetric* angle-ply composites [7-12]. For in this case, despite the fact that $B_{ij} = 0$, the flexural moduli D_{16} and D_{26} do not vanish, which can lead to serious errors in the interpretation of flexure experiments.

7.5 CONCLUSIONS ON CHARACTERIZATION PROCEDURES

For stiffness, creep, and strength characterization of fiber reinforced systems possessing arbitrary fiber orientations and lamination sequences, the experimental procedures outlined here employing thin walled cylindrical test specimens [7-14,16] constitute the most unified procedure available. For the special case of characterization of limited quantities of unidirectional materials, coupon testing of 0° and 90° composites combined with a solid rod torsion experment for shear properties may be employed with the recognition that solid rods often times do not correspond to "equivalent" panels fabricated in most engineering applications.

7.6 MATERIALS PROPERTIES

The following tables and graphs provide data on materials properties for a number of composite materials.

Table 7-1. Dependence of the strength of an orthotropic solid on the sign of the shear stress.

$\theta =$	15°	45°	60°
$+\tau_{xy} =$	3,820 psi	5320	2,690
$-\tau_{xy} =$	1,930 psi	436	1,150

Table 7-3. D. Tabulated data for structure-property calculations.

Material	Epoxy	Polyimide	E-Glass	Boron	Graphite (Thornel-40)
E_{11}	$0.5 X 10^6$ psi	0.4×10^6 psi	10.6×10^6 psi	60×10^6 psi	40×10^6 psi
E_{22}	$0.5 X 10^6$ psi	0.4×10^6 psi	10.6×10^6 psi	60×10^6 psi	1.5×10^6 psi
ν_{12}	.35	.33	.22	0.20	~.2
G_{12}	$.185 \times 10^6$ psi	$.17 \times 10^6$ psi	4.34×10^6	25×10^6 psi	4×10^6 psi
α (in 10^{-6} in/in-°F)	32.0	28-35	2.8	2.8	1.5
K BTU/(hr-ft²-°F/in)	1.7	1.7	7.5		60

Table 7-2. Representative properties for fiber reinforcement.

Fiber	Typical Fiber Diameter 10^{-3} in.	Density	Modulus of Elasticity Tension lb/in² × 10^{-6}	Tensile Strength lb/in²	Expansion Coefficient α 10^{-6} in/in-°F	Thermal Conductivity K BTU/(hr-ft²-°F/in)	Cost (Dollars) Per POund
E Glass	0.4	0.092	10.5	450,000	2.8	7.5	Roving 0.31 Mat 0.44 Cloth 0.80–2.00
S Glass	0.4	0.090	12.3	650,000			3–4.00
970 S Glass	0.4	0.091	14.5	800,000			4.0–4.5
Boron on Tungsten	4	.095	60	400,000	2.8		$250
Graphite	0.2–.4	.053–.065	35–100	250,000–500,000	~1.5	60	350–500
Beryllium	5	.066	45	180,000	6.4	87	10,000
Silicon Carbide on Tungsten	4	0.126	60	360,000	2.2	29	2.500
Stainless Steel	0.5	.283	29	385,000–600,000	~30	100–170	50.00
Asbestos	.001–.01	.090	25	100,000–300,000			1.10
Aluminum	.2	.097	20-60	200,000–300,000	3.7		
Polyamide	.2–.5	.041	0.4	120,000	45.50	1.7	.5–15
Polyester	.8	.052	0.6	100,000	45–50	1.7	.3–1.0

B. STRESS-STRAIN RESPONSE FOR BORON-EPOXY

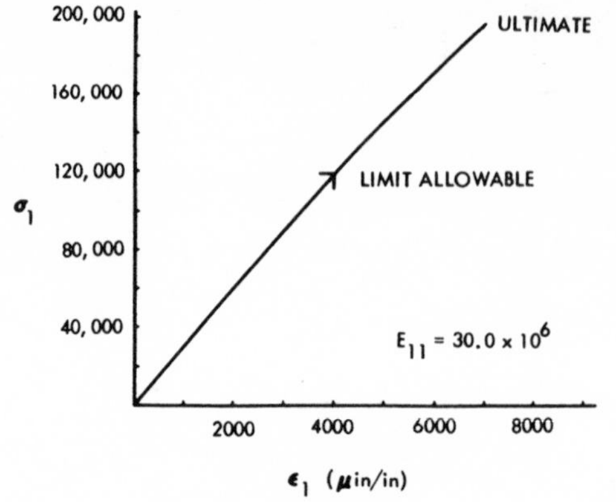

Longitudinal tension.

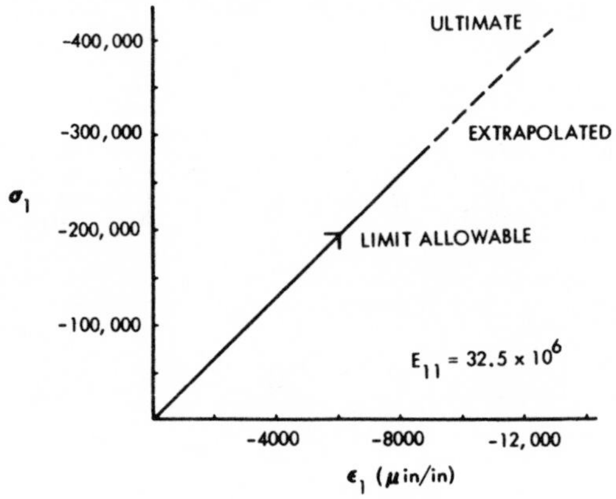

Longitudinal compression.

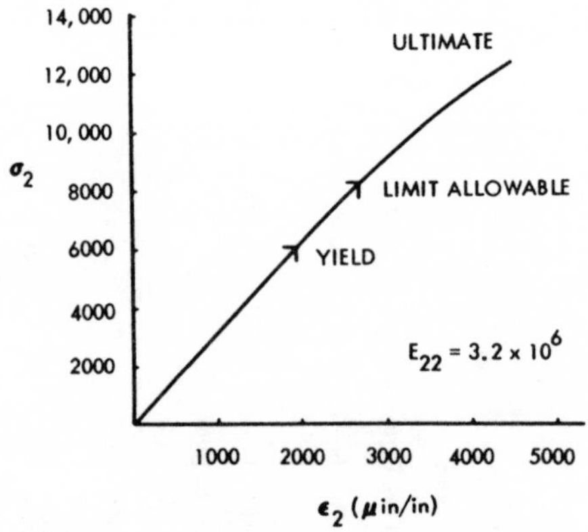

Transverse tension.

B. STRESS-STRAIN RESPONSE FOR BORON-EPOXY

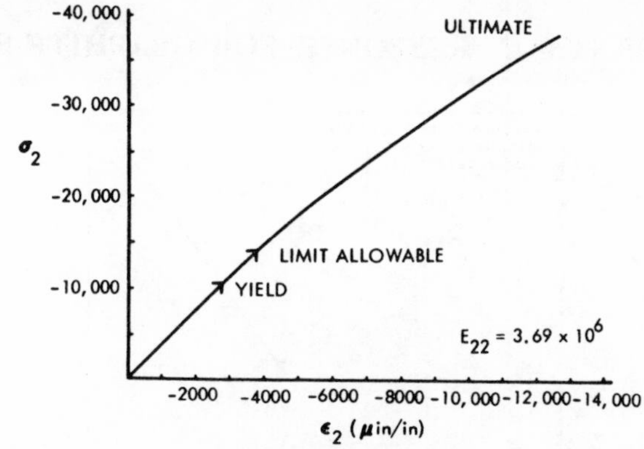

Transverse compression.

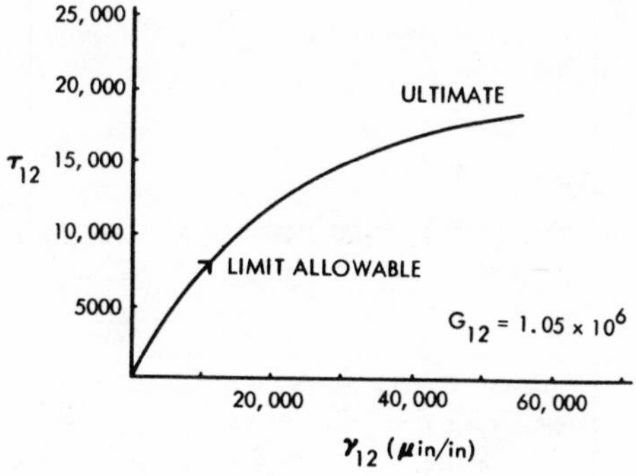

Shear.

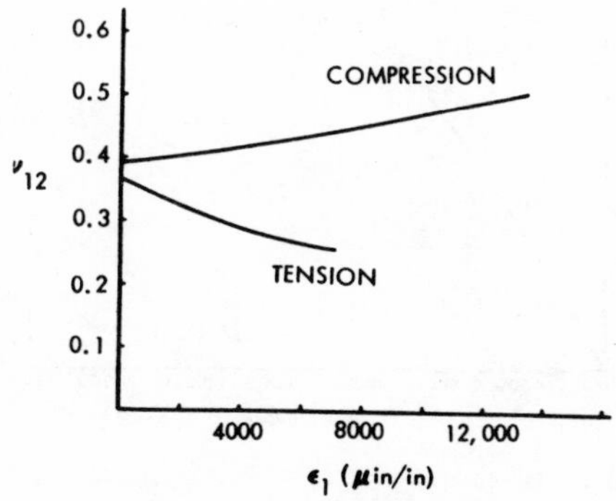

Poisson's ratio versus strain.

C. STRESS-STRAIN RESPONSE FOR GRAPHITE-EPOXY

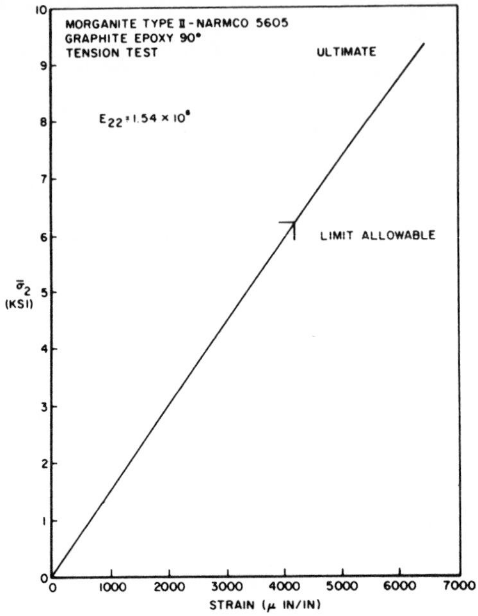

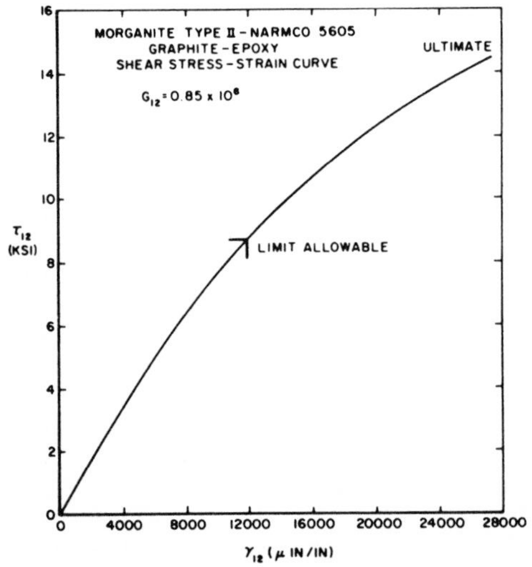

C. STRESS-STRAIN RESPONSE FOR GRAPHITE-EPOXY

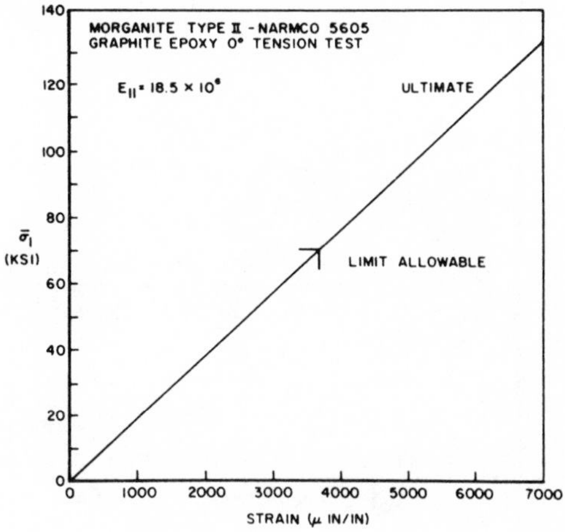

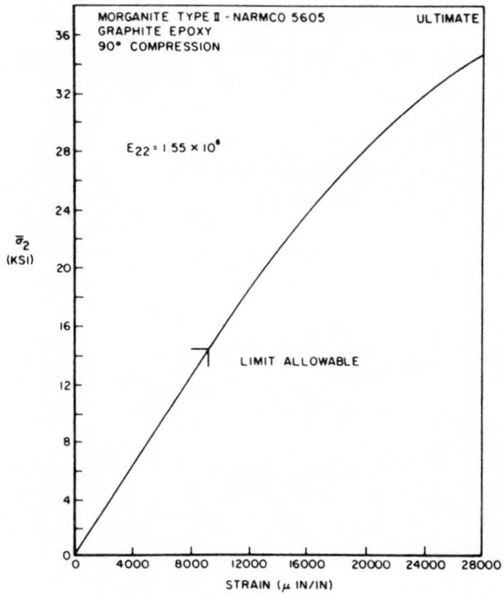

C. STRESS-STRAIN RESPONSE FOR GRAPHITE-EPOXY

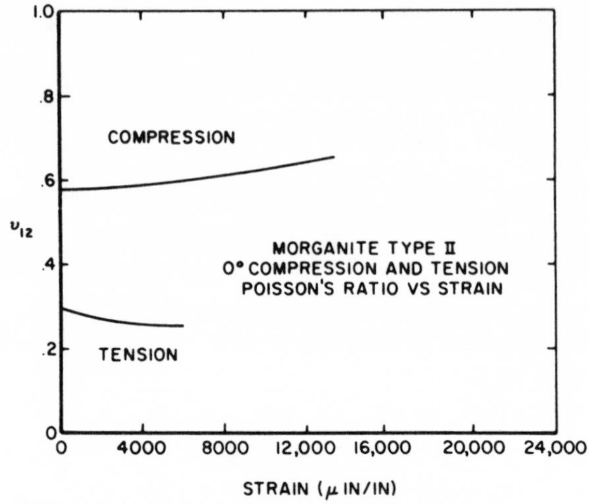

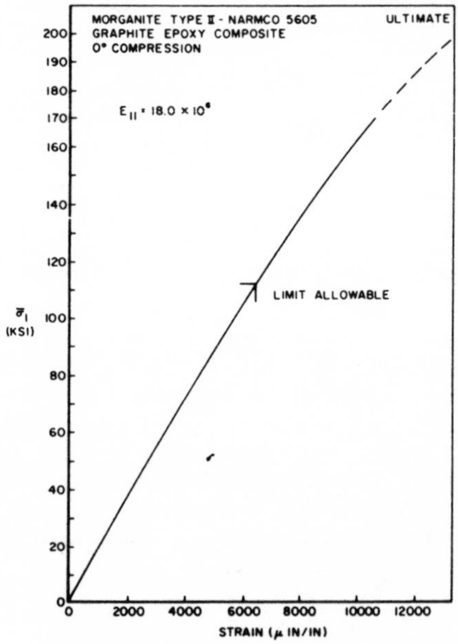

REFERENCES

1. Tsai, Halpin, and Pagano, ed. *Composite Materials Workshop.* Lancaster, PA: Technomic Publishing (1968).
2. Gol'denblat, I. I. and Kapnov, V. A., "Strength of Glass Reinforced Plastics in Complex Stress States," *Mech. Polimerov, 1,* 70 (1965).
3. Halpin, J. C. and Wu, E. M., *J. Comp Mat.* (1969).
4. Wu, E. M. in ref. 1 and worked reported by H. T. Corten in *Fundamental Aspects of Fiber Reinforced Plastic Composites.,* Schwartz and Schwartz, ed. Interscience (1968).
5. Pagano, N. J. and Halpin, J. C., "Influence of End Constraint in the Testing of Anisotropic Bodies," *J. Comp. Mat., 2,* 18 (1968).
6. Kicher, T., Case Institute, unpublished results.
7. Wu, E. M. and Thomas, R. L., "Off-Axis Test of a Composite," *J. Comp. Mat., 2,* 523 (1968).
8. Hoff, N., in *Engineering Laminates,* A. G. H. Dietz, ed. Wiley (1949).
9. Pagano, N. J., "Analysis of the Flexure Test of Bidirectional Composites," *J. Comp. Mat., 1,* 336 (1967).
10. Lekhnitskii, S. G. *Anisotropic Plates.* NY: Gordon and Breach (1968).
11. Hashin, Z., "Plane Anisotropic Beams," *J. Appl. Mech., 34,* 257 (1967).
12. Whitney, J. M., in preparation.
13. Thomas, R. L., Doner, D., and Adams, D., "Mechanical Behavior of Fiber-Reinforced Composite Materials," AFML-TR-67-96 (May 1967).
14. Pagano, N. J., Halpin, J. C., and Whitney, J. M., "Tension Buckling of Anisotropic cylinders," *J. Comp. Mat., 2,* 154 (1968).
15. Dong, S. B., Pister, K. S., and Taylor, R. L., "On the Theory of Laminated Anisotropic Shells and Plates," *J. Aero. Sci.,* 969 (1962).
16. Whitney, J. M. and Halpin, J. C., "Analysis of Laminated Anisotropic Tubes Under Combined Loading," *J. Comp. Mat., 2,* 360 (1968).
17. Whitney, J. M., "Engineering Constants of Laminated Composite Materials," *J. Comp. Mat., 2,* 261 (1968).

APPENDIX A

Matrices and Tensors

THE MATHEMATICS OF MATRICES AND OF TENSORS GIVES THE ENGINEER or scientist an additional mathematical tool and a shorthand method of expressing himself. Matrix and tensor notation has been utilized quite extensively in the literature on laminated composites; therefore, it is felt that a brief and concise presentation of matrix algebra and tensor concepts is in order.

MATRIX ALGEBRA

A matrix is a rectangular array of *i* rows and *j* columns of numbers or quantities such as:

$$Q = Q_{ij} = \begin{bmatrix} Q_{11} & Q_{12} & Q_{13} \\ Q_{21} & Q_{22} & Q_{23} \end{bmatrix} \quad \begin{matrix} i = 1,2 \\ j = 1,2,3 \end{matrix}$$

or, for a square matrix

$$Q = Q_{ij} = \begin{bmatrix} Q_{11} & Q_{12} & Q_{13} \\ Q_{21} & Q_{22} & Q_{23} \\ Q_{31} & Q_{32} & Q_{33} \end{bmatrix} \quad \begin{matrix} i = 1,2,3 \\ j = 1,2,3 \end{matrix}$$

A row or a column matrix can be expressed as

$$Q = Q_1 = \begin{bmatrix} Q_1 \\ Q_2 \\ Q_3 \end{bmatrix} \quad i = 1,2,3$$

$$Q = Q_j = [Q_1 \ \ Q_2 \ \ Q_3] \quad j = 1,2,3$$

The row or column matrix is of the same form as a vector. For instance, the velocity vector can be expressed as:

$$V = \begin{bmatrix} V_x \\ V_y \\ V_z \end{bmatrix}$$

The SUM of two matrices, *A* and *B*, is defined only when *A* and *B* have the same number of rows and columns. The sum is obtained by adding corresponding elements of A and B. In other words,

$$C_{ij} = A_{ij} + B_{ij} \tag{A-1}$$

or, if $i = j = 1,2$

$$C_{11} = A_{11} + B_{11} \quad C_{21} = A_{21} + B_{21}$$
$$C_{12} = A_{12} + B_{12} \quad C_{22} = A_{22} + B_{22}$$

Therefore, the matrix, C, is given by

$$\begin{bmatrix} C_{11} & C_{12} \\ C_{21} & C_{22} \end{bmatrix} = \begin{bmatrix} A_{11} & A_{12} \\ A_{21} & A_{22} \end{bmatrix} + \begin{bmatrix} B_{11} & B_{12} \\ B_{21} & B_{22} \end{bmatrix} = \begin{bmatrix} (A_{11} + B_{11}) & (A_{12} + B_{12}) \\ (A_{21} + B_{21}) & (A_{22} + B_{22}) \end{bmatrix}$$

The sum of two matrices is a commutative operation. In other words,

$$A + B = B + A$$

Also, the sum of two matrices is an associative operation. That is,

$$(A + B) + C = A + (B + C)$$

The DIFFERENCE of two matrices is defined in a manner similar to that for the sum. The difference is obtained by subtracting corresponding elements of the matrices:

$$C_{ij} = A_{ij} - B_{ij} \tag{A-2}$$

or, when $i = j = 1,2$

$$C_{11} = A_{11} - B_{11}$$
$$C_{12} = A_{12} - B_{12}$$
$$C_{21} = A_{21} - B_{21}$$
$$C_{22} = A_{22} - B_{22}$$

The difference of two matrices is also a communicative and associative operation.

The PRODUCT of a scalar, s, and a Matrix, A, is obtained by multiplying each element of A by the scalar. In other words, when

$$A = \begin{bmatrix} A_{11} & A_{12} \\ A_{21} & A_{22} \end{bmatrix}$$

then

$$sA = \begin{bmatrix} sA_{11} & sA_{12} \\ sA_{21} & sA_{22} \end{bmatrix} \tag{A-3}$$

The PRODUCT of two matrices, A and B, is defined only for the case where the number of rows of B equals the number of columns in A. When this is the case, the product AB is given by the matrix C which is obtained by MULTIPLYING each element of the ith row of A by the corresponding element of the jth column of B and ADDING. In other words,

$$\begin{bmatrix} C_{11} \\ C_{21} \\ C_{31} \end{bmatrix} = \begin{bmatrix} A_{11} & A_{12} & A_{13} \\ A_{21} & A_{22} & A_{23} \\ A_{31} & A_{32} & A_{33} \end{bmatrix} \begin{bmatrix} B_{11} \\ B_{21} \\ B_{31} \end{bmatrix} \tag{A-4}$$

where

$$\begin{aligned} C_{11} &= A_{11}B_{11} + A_{12}B_{21} + A_{13}B_{31} \\ C_{21} &= A_{21}B_{11} + A_{22}B_{21} + A_{23}B_{31} \\ C_{31} &= A_{31}B_{11} + A_{32}B_{21} + A_{33}B_{31} \end{aligned} \tag{A-5}$$

Note that if "m" is the number of rows in the A matrix and "n" is the number of columns in the B matrix, then the C matrix will have m rows and n columns. Note that equations (A-4) and (A-5) are similar in nature to the lamina and laminate constitutive equations which were presented in matrix form in the text. Of course, the subscript notation for the constitutive equations has been changed slightly to account for the symmetry of the Hooke's law matrix.

Note that in general, premultiplying a matrix B by a matrix A does not give the same as postmultiplying. In other words,

$$AB \neq BA$$

The TRANSPOSE of a matrix A is defined as the matrix, A^T, which has the rows and columns of A interchanged. In other words, the transpose of the Q matrix

$$Q = \begin{bmatrix} Q_{11} & Q_{12} \\ Q_{21} & Q_{22} \end{bmatrix}$$

is given by

$$Q^T = \begin{bmatrix} Q_{11} & Q_{21} \\ Q_{12} & Q_{22} \end{bmatrix} \tag{A-6}$$

A UNIT MATRIX is defined as an m by m square matrix where the diagonal elements equal unity and the off diagonal elements are zero. In other words,

$$E = \begin{bmatrix} 1 & 0 \\ 0 & 1 \end{bmatrix}$$

or (A-7)

$$E = \begin{bmatrix} 1 & 0 & 0 \\ 0 & 1 & 0 \\ 0 & 0 & 1 \end{bmatrix}$$

The COFACTOR matrix of an m by m square matrix is defined as the matrix which is obtained by replacing each element of the A matrix by its cofactor. The cofactor of the element is defined as the product of the determinant of the matrix with m-1 rows and m-1 columns, which is obtained by erasing the ith row and j^{th} column of A, by the term $(-1)^{i+j}$. As an example, the cofactor matrix of

$$A = \begin{bmatrix} 3 & 4 & 1 \\ 2 & 1 & 6 \\ 5 & 3 & 2 \end{bmatrix} \tag{A-8}$$

is determined as follows:

$$CoA_{11} = \begin{vmatrix} 1 & 6 \\ 3 & 2 \end{vmatrix} (-1)^{(1+1)} = (2 - 18)(1)^2 = -16$$

$$CoA_{12} = \begin{vmatrix} 2 & 6 \\ 3 & 2 \end{vmatrix} (-1)^{(1+2)} = (4 - 30)(-1)^3 = 26$$

$$CoA_{13} = \begin{vmatrix} 2 & 1 \\ 5 & 3 \end{vmatrix} (-1)^{(1+3)} = (6 - 5)(-1)^4 = 1$$

$$CoA_{21} = \begin{vmatrix} 4 & 1 \\ 3 & 2 \end{vmatrix} (-1)^{(2+1)} = (8 - 3)(-1)^3 = -5$$

etc.

therefore, the cofactor matrix of A is given by

$$CoA = \begin{bmatrix} -16 & 26 & 1 \\ -5 & 1 & 11 \\ 22 & -16 & -5 \end{bmatrix}$$

The transpose of a matrix, cofactor of a matrix, and the unit matrix will be used in the definition of one of the most useful concepts in matrix algebra—the INVERSE matrix. For certain square matrices, A, the INVERSE matrix, A^{-1}, can be defined as

$$A A^{-1} = E \tag{A-9}$$

where E is a unit matrix and the INVERSE matrix A^{-1}, is given by

$$A^{-1} = \frac{(CoA)^T}{|A|} \tag{A-10}$$

In other words, the inverse matrix is given by the transpose of the cofactor matrix divided by the determinant. As an example, the INVERSE of the matrix given in equation (A-8) is computed as follows:

$$[A]^{-1} = \frac{\begin{bmatrix} -16 & -5 & 22 \\ 26 & 1 & -16 \\ 1 & 11 & 5 \end{bmatrix}}{\begin{vmatrix} 3 & 4 & 1 \\ 2 & 1 & 6 \\ 5 & 3 & 2 \end{vmatrix}} = \frac{\begin{bmatrix} -16 & -5 & 22 \\ 26 & 1 & -16 \\ 1 & 11 & 5 \end{bmatrix}}{57}$$

Therefore,

$$[A]^{-1} = \begin{bmatrix} -\frac{16}{57} & -\frac{5}{57} & \frac{22}{57} \\ \frac{26}{57} & \frac{1}{57} & -\frac{16}{57} \\ \frac{1}{57} & \frac{11}{57} & \frac{5}{57} \end{bmatrix}$$

It may be easily verified by matrix multiplication that

$$[A][A]^{-1} = \begin{bmatrix} 1 & 0 & 0 \\ 0 & 1 & 0 \\ 0 & 0 & 1 \end{bmatrix}$$

TENSORS

A TENSOR is a physical entity in nature which obeys certain transformation relations. There are different ranks of tensors which correspond to different degrees of complexity of the tensor. Each rank of tensors have their own transformation relations. It is only necessary to establish that a physical entity is a tensor and determine its rank, and its transformation relations are known. Most of the familiar "engineering quantities" are tensors. Some of the familiar tensors are speed, velocity, strain, and the Hooke's law relations (stiffnesses). Although tensors are expressed in a form similar to matrices, a matrix is not necessarily a tensor. The components of the matrix must obey the tensor transformation relations before the matrix may be classified a tensor.

The simplest tensor is a SCALAR, and it is a tensor of zero-th rank. A scalar is an entity which has only magnitude not direction. No transformation relations are required for scalars since they have only magnitude. An example of a scalar is speed.

A VECTOR happens to be a tensor of 1st rank, and it has magnitude and direction associated with its components. It is known that the components of a vector change as the coordinate system is rotated, and the change which occurs in the components of the vector is governed by certain mathematical relations. These are the transformation relations for vectors or 1st rank tensors. For instance, if one desired to transform the components of a vector, V, from one set of coordinates, (x,y) to a second set of coordinates $(1,2)$ by a rotation about the $(3,z)$ axes, the following transformation matrix would be used.

$$[T] = \begin{bmatrix} \cos\theta & \sin\theta & 0 \\ -\sin\theta & \cos\theta & 0 \\ 0 & 0 & 1 \end{bmatrix} \tag{A-11}$$

or, with reference to Figure A-1, the transformation would be

$$\begin{bmatrix} V_1 \\ V_2 \\ V_3 \end{bmatrix} = \begin{bmatrix} \cos\theta & \sin\theta & 0 \\ -\sin\theta & \cos\theta & 0 \\ 0 & 0 & 1 \end{bmatrix} \begin{bmatrix} V_x \\ V_y \\ V_z \end{bmatrix}$$

Thus, equation (A-11) is the transformation relations for a 1st rank tensor (vector) for a rotation about the 3 or Z axes.

As it turns out, stress and strain are both second rank tensors. They obey the transformation relations for second rank tensors. These transformations are well known, and the graphical form of these transformation relations is the familiar Mohr's Circle. These transformation relations are derived in

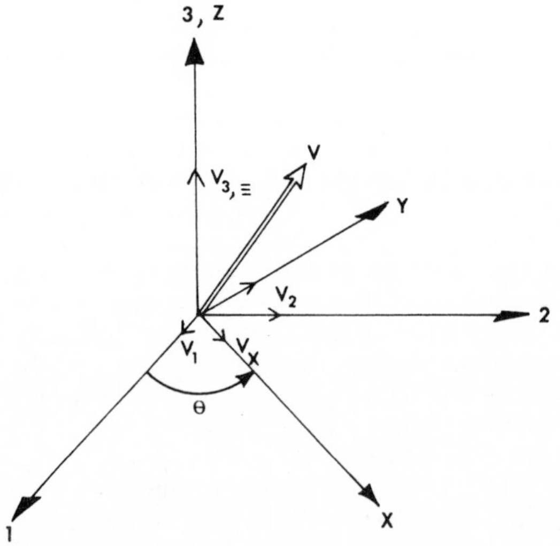

Figure A-1. *Coordinate transformation.*

Chapter 2 for stress and strain. The relations are presented here in order that a comparison can be made with the relations presented for 1st rank tensors, equation (A-11).

$$[T] = \begin{bmatrix} \cos^2\theta & \sin^2\theta & 2\sin\theta\cos\theta \\ \sin^2\theta & \cos^2\theta & -2\sin\theta\cos\theta \\ -\sin\theta\cos\theta & \sin\theta\cos\theta & (\cos^2\theta - \sin^2\theta) \end{bmatrix} \qquad \text{(A-12)}$$

A comparison of the transformations given by equation (A-12) shows that they are more complex than the transformations for 1st rank tensors (vectors).

The transformation relations would be utilized to transform plane stress and strain states from one set of axes *(1,2)* to a second set of axes *(x,y)* through a rotation about the *(3,z)* axes as follows:

$$\begin{bmatrix} \sigma_1 \\ \sigma_2 \\ \tau_{12} \end{bmatrix} = [T] \begin{bmatrix} \sigma_x \\ \sigma_y \\ \tau_{xy} \end{bmatrix} \qquad \text{(A-13)}$$

and

$$\begin{bmatrix} \varepsilon_1 \\ \varepsilon_2 \\ \dfrac{\gamma_{12}}{2} \end{bmatrix} = [T] \begin{bmatrix} \varepsilon_x \\ \varepsilon_y \\ \dfrac{\gamma_{xy}}{2} \end{bmatrix} \qquad \text{(A-14)}$$

Certain material properties, such as thermal coefficient of expansion, are also second rank tensors if the material is not isotropic. If the material happens to be isotropic, these properties are tensors of the zero-th rank (scalars), and no transformations are required since the properties are the same in all directions. If the material happens to be ORTHOTROPIC, the transformations in the *(1,2)* or *(X, Y)* plane for the thermal coefficient of expansion are as follows:

$$\begin{bmatrix} \alpha_1 \\ \alpha_2 \\ 0 \end{bmatrix} = [T] \begin{bmatrix} \alpha_{xy} \\ \alpha_y \\ \dfrac{1}{2}\alpha_{xy} \end{bmatrix} \qquad \text{(A-15)}$$

The Hooke's law relations or stiffness of a material are components of a fourth rank tensor. The transformation relations for a specially orthotropic material under a rotation about the *(3,Z)* axes are derived in Chapter 2. The relations are presented here in matrix form:

$$\begin{bmatrix} Q_{11} \\ Q_{22} \\ Q_{12} \\ Q_{66} \\ Q_{16} \\ Q_{26} \end{bmatrix} = \begin{bmatrix} m^4 \\ n^4 \\ m^2n^2 \\ m^2n^2 \\ -m^3n \\ -mn^3 \end{bmatrix} \begin{bmatrix} n^4 & 2m^2n^2 & 4m^2n^2 \\ m^4 & 2m^2n^2 & 4m^2n^2 \\ m^2n^2 & (m^4+n^4) & -4m^2n^2 \\ m^2n^2 & -2m^2n^2 & (m^2-n^2)^2 \\ nm^3 & (m^3n-mn^3) & 2(m^3n-mn^3) \\ m^3n & (mn^3-m^3n) & 2(mn^3-m^3n) \end{bmatrix} \begin{bmatrix} Q_{11} \\ Q_{22} \\ Q_{12} \\ Q_{66} \\ 0 \\ 0 \end{bmatrix} \tag{A-16}$$

$$m = \cos\theta$$
$$n = \sin\theta$$

It is evident that these transformation relations are more complex than those for the first or second rank tensors. Although these relations are more complex, they represent the same type of transformations as the familiar Mohr's Circle transformations for stress and strain. This transformation of material stiffnesses is probably a new concept since it is unnecessary for isotropic materials. Note that equation (A-16) accounts for rotations with respect to the material coordinate system in contrast to equations (2-35) which are with respect to the x,y coordinate system.

A P P E N D I X B

Equilibrium Equations for Plates

THE EQUATIONS OF EQUILIBRIUM CAN BE OBTAINED BY CONSIDERING AN infinitesimal element cut from the plate, Figure B-1. This element has a thickness h, but this thickness can be disregarded by considering the stress system in terms of the stress and moment resultants (defined in Chapter 3) acting at the geometrical midplane. These stress resultants are shown in Figure B-1a and the moment resultants in Figure B-1b.

In addition to the stress resultants N_x, N_y, and N_{xy} defined in Chapter 3, two transverse shear stress resultants are shown in Figure B-1c. These resultants are defined in a manner annalogous to the definitions for N_x, N_y, and N_{xy}:

$$Q_x = \int_{-h/2}^{h/2} \tau_{xz}\, dz \tag{B-1}$$

$$Q_y = \int_{-h/2}^{h/2} \tau_{yz}\, dz \tag{B-2}$$

That is, the transverse shear stress resultants Q_x and Q_y are the sum or integral of the transverse shears across the thickness. They have dimensions of force/length. These transverse shear stress resultants are shown in Figure B-1c along with the distributed vertical loading $q\,(x,y)$. The actual stress system acting on the element is composed of the sum of the stress resultants N_x, N_y, N_{xy}, Q_x, Q_y and the moment resultants M_x, M_y, M_{xy}. This system must be in equilibrium with the applied distributed load $q\,(x,y)$. To insure this, we must have equilibrium with respect to forces in the x, y, and z directions, as well as moment equilibrium with respect to the x and y axes. These conditions give us five equilibrium equations in terms of the rates of change of the stress and moment resultants. That is, we consider the rate of change of a quantity in the appropriate direction, and consider the change over the elemental length. Thus on the edge $x = 0$ N_x acts, and since the rate of change is $\partial N_x/\partial x$, this stress resultant has the value $N_x + (\partial N_x/\partial x)\, dx$ after the length dx, see Figure B-1a. Below we consider each of the five equilibrium conditions:
Summation of forces in the x direction = 0:

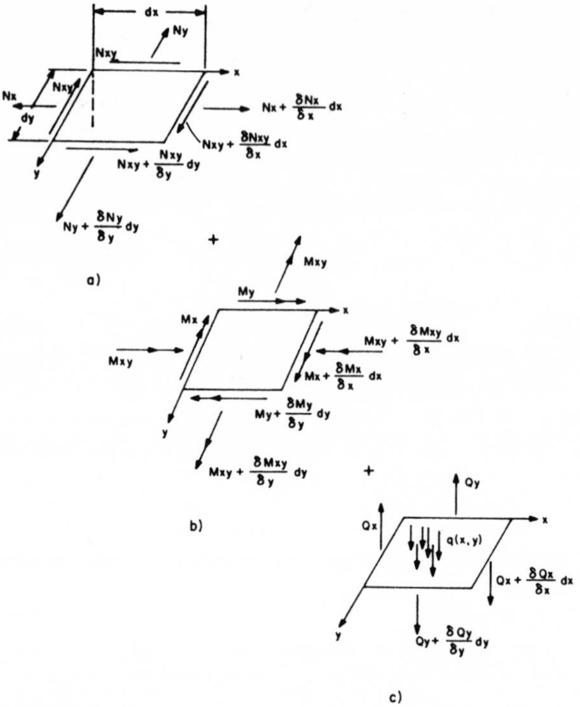

Figure B-1. *Midplane stress and moment resultant system.*

$$N_x dy + \frac{\partial N_x}{\partial x}\, dy + N_{xy} dx + \frac{\partial N_{xy}}{\partial y}\, dy\, dx - N_x dy - N_{xy} dx = 0$$

or

$$\frac{\partial N_x}{\partial x} + \frac{\partial N_{xy}}{\partial y} = 0 \qquad \text{(B-3)}$$

Summation of forces in the y direction = 0:

$$N_y dy + \frac{\partial N_y}{\partial x}\, dy\, dx + N_{xy} dy + \frac{\partial N_{xy}}{\partial y}\, dx\, dy - N_y dx - N_{xy} dy = 0$$

or

$$\frac{\partial N_y}{\partial y} + \frac{\partial N_{xy}}{\partial x} = 0 \qquad \text{(B-4)}$$

Summation of forces in the z direction = 0:

$$Q_x dy + \frac{\partial Q_x}{\partial x} dx\, dy + Q_y dx + \frac{\partial Q_y}{\partial y} dy\, dx - Q_x dy - Q_y dx + q(x,y) dx\, dy = 0$$

or

$$\frac{\partial Q_x}{\partial x} + \frac{\partial Q_y}{\partial y} + q(x,y) = 0 \qquad \text{(B-5)}$$

Summation of moments about the x axis = 0:

$$- M_y dx - \frac{\partial M_y}{\partial y} dy\, dx - M_{xy} dy - \frac{\partial M_{xy}}{\partial x} dx\, dy + Q_y dx\, dy + \frac{\partial Q_y}{\partial y} dy\, dx\, dy$$

$$+ \; q(x,y) dx\, dy\, dy/2 + Q_x dy\, dy/2 + \frac{\partial Q_x}{\partial x} dx\, dy\, dy/2$$

$$+ \; M_y dx + M_{xy} dy - Q_x dy\, dy/2 = 0$$

or collecting and dropping the higher order terms (since for arbitrarily small dx and dy these terms vanish):

$$- \frac{\partial M_y}{\partial y} - \frac{\partial M_{xy}}{\partial x} + Q_y = 0$$

or

$$Q_y = \frac{\partial M_y}{\partial y} + \frac{\partial M_{xy}}{\partial x} \qquad \text{(B-6)}$$

Summation of moments about the y axis = 0:

$$M_x dy \; + \frac{\partial M_x}{\partial x} dx\, dy + M_{xy} dx + \frac{\partial M_{xy}}{\partial y} dy\, dx - Q_x dy\, dx$$

$$- \frac{\partial Q_x}{\partial x} dx\, dy\, dx - Q_y dx\, dx/2 - \frac{\partial Q_y}{\partial y} dy\, dx\, dx/2 + q(x,y) dx\, dy\, dx/2$$

$$- \; M_x dy - M_{xy} dx + Q_y dx\, dx/2 = 0$$

or collecting and dropping the higher order terms:

$$\frac{\partial M_x}{\partial x} + \frac{\partial M_{xy}}{\partial y} = Q_x \qquad \text{(B-7)}$$

Equations (B-3) through (B-7) express the five relevant conditions of equilibrium for the plate element. The last three can be combined to eliminate the transverse shear stress resultants, and yield a single moment equilibrium equation:

$$\frac{\partial}{\partial x}\left(\frac{\partial M_x}{\partial x} + \frac{\partial M_{xy}}{\partial y}\right) + \frac{\partial}{\partial y}\left(\frac{\partial M_y}{\partial y} + \frac{\partial M_{xy}}{\partial x}\right) = -\, q(x,y)$$

or

$$\frac{\partial^2 M_x}{\partial x^2} + 2\,\frac{\partial^2 M_{xy}}{\partial x \partial y} + \frac{\partial^2 M_y}{\partial y^2} = -\, q(x,y) \qquad \text{(B-8)}$$

A P P E N D I X C

Trigometric Functions for Laminated Composites

θ	sin 2θ	Cos 2θ	sin 4θ	Cos 4θ
0	0	1	0	1
15	1/2	0.866	0.866	1/2
30	0.866	1/2	0.866	-1/2
45	1.0	0	0	-1
60	0.866	-1/2	-0.866	-1/2
75	1/2	-0.866	-0.866	1/2
90	0	-1	0	1